Uncertainty in Wastewater Treatment Design and Operation

Addressing Current Practices and Future Directions

Scientific and Technical Report Series No. 21

Uncertainty in Wastewater Treatment Design and Operation

Addressing Current Practices and Future Directions

Edited by

Evangelia Belia, Lorenzo Benedetti, Bruce Johnson, Sudhir Murthy, Marc Neumann, Peter Vanrolleghem and Stefan Weijers

IWA PUBLISHING

Published by **IWA Publishing**
Unit 104–105, Export Building
1 Clove Crescent
London E14 2BA, UK
Telephone: +44 (0)20 7654 5500
Fax: +44 (0)20 7654 5555
Email: publications@iwap.co.uk
Web: www.iwapublishing.com

First published 2021

British Library Cataloguing in Publication Data
A CIP catalogue record for this book is available from the British Library

ISBN: 9781780401027 (Paperback)
ISBN: 9781780401034 (eBook)
ISBN: 9781789062601 (ePUB)

Contents

List of Contributors

TASK GROUP MEMBERS

Evangelia Belia, Primodal Inc., US&Canada
Lorenzo Benedetti, Waterways d.o.o., Croatia
Bruce Johnson, Jacobs, USA
Sudhir Murthy, NEWhub Corp., USA
Marc Neumann, Basque Centre for Climate Change, Spain
Peter Vanrolleghem, Université Laval, Canada
Stefan Weijers, Waterschap De Dommel, Netherlands

STR CHAPTER EVOLUTION AND CONTRIBUTING AUTHORS

The STR materials originally consisted of white papers with contributions from a number of authors as shown in Table 1.

Table 1. Original STR white papers.

Task A0	Document how uncertainty and risk are currently handled in the wastewater treatment practice by consultants, utilities and regulators. Brian Karmasin, Henryk Melcer, Jeffrey McCormick, Ana Pena-Tijerina, Davide Bixio, Jose Jimenez
Task A1	Propose a set of terms and definitions relating to uncertainty. Kris Villez, Dave Kinnear, Sylvie Gillot
Task A2	Propose a comprehensive list of the sources of uncertainty for typical project planning horizons. Marie Burbano, Charles Bott, JB Neethling, Maureen O'Shaughnessy, Leiv Rieger, Youri Amerlinck, Benedek Plosz
Task A3	Document and evaluate existing methods for assessing and evaluating uncertainty in wastewater treatment. George Sprouse, Gurkan Sin, Jeffrey McCormick, Oliver Schraa
Task A4	Identify gaps and inefficiencies in current knowledge and practice related to uncertainty. Andrew Shaw, Thomas Hug, Jeffrey McCormick, Youri Amerlinck, Xavier Flores-Alsina, Jeffrey Weiss

Task A5	Incorporate knowledge from other fields and institutions on applications of uncertainty evaluation methodologies. Oliver Schraa, Kris Villez, Spencer Snowling, Benedek Plosz
Task A6	Finalise report. Evangelia Belia, Marc Neumann, Lorenzo Benedetti, Bruce Johnson, Sudhir Murthy, Stefan Weijers, Peter Vanrolleghem

The final version of the STR, following four sets of reviews, is shown in Table 2. In the process of stream lining the material, the content of several of the white papers was re-organised and distributed in different chapters.

Table 2. Final version of STR chapters.

Chapter 1	Key Concepts Marc Neumann and Evangelia Belia
Chapter 2	Current Practice Brian Karmasin, Henryk Melcer, Jeffrey McCormick, Ana Pena-Tijerina, Davide Bixio, Jose Jimenez, Bruce Johnson
Chapter 3	Benefits Marc Neumann, Andrew Shaw, Thomas Hug, Jeffrey McCormick, Youri Amerlinck, Xavier Flores-Alsina
Chapter 4	Evaluation of Existing Methods George Sprouse, Gurkan Sin, Jeffrey McCormick, Oliver Schraa
Chapter 5	DOUT Framework Evangelia Belia, Marie Burbano, Charles Bott, JB Neethling, Maureen O'Shaughnessy, Leiv Rieger, Youri Amerlinck, Benedek Plosz, Peter Vanroleghem
Chapter 6	Case Studies Evangelia Belia, Bruce Johnson, Peter Vanroleghem
Chapter 7	Bigger Picture Evangelia Belia, Lorenzo Benedetti, Bruce Johnson, Sudhir Murthy, Marc Neumann, Peter Vanrolleghem, Stefan Weijers
Chapter 8	Future of Uncertainty Analysis Marc Neumann, Andrew Shaw, Thomas Hug, Jeffrey McCormick, Youri Amerlinck, Xavier Flores-Alsina, Jeffrey Weiss
Appendix A	Terminology and Definitions Kris Villez, Dave Kinnear, Sylvie Gillot
Appendix B	Theory of Methods Oliver Schraa, Kris Villez, Spencer Snowling, Benedek Plosz
Appendix C	Literature Search Results George Sprouse, Gurkan Sin, Jeffrey McCormick, Oliver Schraa, Lorenzo Benedetti
Appendix D	Methods from Other Fields Oliver Schraa, Kris Villez, Spencer Snowling, Benedek Plosz
Appendix E	Current Practice Worldwide Evangelia Belia, John Copp, Bruce Johnson, Marc Neumann, Libor Novák, Daniel Nolasco, Dae Sung Lee, Stefan Weijers

CONTRIBUTING AUTHORS

Evangelia Belia
Primodal Inc., US&Canada

Lorenzo Benedetti
Waterways d.o.o., Croatia

Bruce Johnson
Jacobs, USA

Sudhir Murthy
NEWhub Corp, USA

Marc Neumann
Basque Centre for Climate Change (BC3)
Ikerbasque – Basque Foundation for Science, Spain

Peter Vanrolleghem
modelEAU-Université Laval, Canada

Stefan Weijers
Waterschap De Dommel, Netherlands

Youri Amerlinck
Aquafin NV, Belgium

Davide Bixio
European Commission, Belgium

Charles Bott
Hampton Roads Sanitation District, USA

Marie Burbano
Brown and Caldwell, USA

Xavier Flores-Alsina
Technical University of Denmark, Denmark

Sylvie Gillot
INRAE, France

Thomas Hug
Hunziker Betatech AG, Switzerland

Jose Jimenez
Brown and Caldwell, USA

Brian Karmasin
CDM Smith, USA

David Kinnear
Kinnear Process Solutions LLC, USA

Jeffrey McCormick
Carollo Engineers, USA

Henryk Melcer
Brown and Caldwell, USA

JB Neethling
HDR, USA

Daniel Nolasco
NOLASCO y Asociados S.A., Argentina

Libor Novák
aqua4you s.r.o., Czech Republic

Maureen O'Shaughnessy
Prince William County Service Authority, USA

Anna Pena-Tijerina
Plummer Associates Inc., USA

Leiv Rieger
inCTRL Solutions Inc., Canada

Oliver Schraa
inCTRL Solutions Inc., Canada

Andrew Shaw
Black & Veatch, USA

Gurkan Sin
Technical University of Denmark, Denmark

Spencer Snowling
Hydromantis Inc., Canada

George Sprouse
MCES, USA

Benedek Plosz
University of Bath, UK

Kris Villez
Oak Ridge National Laboratory, USA

Jeff Weiss
Sanitation Districts of Los Angeles County (retired),USA

Preface

ABOUT THE IWA DESIGN AND OPERATIONAL UNCERTAINTY TASK GROUP

The International Water Association (IWA) Design and Operations Uncertainty Task Group (DOUT) was formed to develop methodologies that enable the explicit evaluation of variability and uncertainty in model-based design of water resource recovery facilities (WRRF), and model-based analysis of plant operations.

An overview of uncertainty in the treatment plant modelling context was discussed at a workshop (in Mont-Sainte-Anne, Canada) during the 1st IWA/WEF (Water Environment Federation) Wastewater Treatment Modelling seminar (WWTmod2008). This workshop identified knowledge gaps and the requirements for the development of the needed methodologies. Following the workshop, the Task Group established the following set of objectives and set-up several working groups to advance these goals:

- Document how uncertainty and risk are currently handled in wastewater treatment practice by consultants, utilities and regulators.
- Propose a set of terms and definitions relating to uncertainty to be used by wastewater professionals.
- Propose a comprehensive list of the sources of uncertainty for typical project phases and contract delivery mechanisms.
- Document and evaluate existing methods for assessing and evaluating uncertainty in wastewater treatment.
- Identify gaps and inefficiencies in current knowledge and practice related to uncertainty.
- Incorporate uncertainty evaluation methodology knowledge from other fields.
- Present examples of methods already available that can be used to deal with uncertainty and variability.

Editor(s): Evangelia Belia, Lorenzo Benedetti, Bruce Johnson, Sudhir Murthy, Marc Neumann, Peter Vanrolleghem and Stefan Weijers
doi: 10.2166/9781780401034_xv

The working groups were composed of professionals from consulting, utilities, software companies and academia. From its inception the intention was one of ***co-production***, further facilitated through a large number of workshops and working meetings held during national and international conferences. The findings obtained through this process form the cornerstone of this Scientific and Technical Report (STR).

MISSION STATEMENT

The goal of the Task Group was to develop methods for integrating uncertainty analysis into wastewater treatment process simulators in order to facilitate a shift from deterministic (one answer) to probabilistic analysis (likelihood of outcome) of treatment plant design and operation. Such a transition will lead to better management and quantification of the risks/benefits of a specific design or operational strategy. This in turn will provide utilities with more effective, efficient facilities and increase the socio-economic benefits of resource recovery.

In pursuit of these objectives, this STR reviews the state of the art in dealing with uncertainty and variability in wastewater engineering, as well as novel methods and approaches recently developed in academia. The STR examines the feasibility of these novel methods for use in the wastewater sector.

SCOPE

The work presented in the STR, focuses on the entire wastewater treatment plant from influent to effluent. Links to the urban catchment (upstream of the wastewater treatment plant) are also discussed because uncertainties associated with expected developments in the catchment have impacts at the planning stage of plant design. Links to the receiving water body (downstream of the wastewater treatment plant) are also discussed as uncertainties in effluent standards imposed by regulators impact plant design and operation.

Much of the work presented in this STR focuses on biological treatment in the liquid stream as this is one of the principal drivers for initiating this paradigm change in design methodology. However, it is important to note that the methodologies presented are model-independent and applicable to any unit process (e.g., primary settling tank, anaerobic digester, etc.), including external factors, or even within an all-encompassing plant-wide modelling approach.

The Task Group hopes that the concepts and methods presented in this STR will contribute to a more systematic and transparent way of managing uncertainty in WRRF design and operations, which in turn will lead to more cost-effective solutions.

Evangelia Belia, Canada
Lorenzo Benedetti, Croatia
Bruce Johnson, USA
Sudhir Murthy, USA
Marc Neumann, Spain
Peter Vanrolleghem, Canada
Stefan Weijers, Netherlands

Acknowledgements

The Design and Operations Uncertainty Task Group (DOUT) was sponsored by the International Water Association (IWA). We would especially like to thank Michael Dunn, former managing director at IWA Publishing and Paul Reiter, former IWA executive director, as well as Mark Hammond and Niall Cunniffe for guiding us through the publishing effort.

The Task Group was also supported by the Water Environment Federation (WEF) and we would like to thank them for their support.

The Natural Sciences and Engineering Research Council of Canada (NSERC) and Primodal Inc., funded a Collaborative Research and Development grant which supported the development of the STR.

The following organisations provided financial support through the sponsoring of projects:

- Waterschap de Dommel, The Netherlands
- Water Research Foundation, Virginia, USA
- Hampton Roads Sanitation District, Virginia, USA

The doctoral research of Mansour Talebizadeh, co-supervised by Cristina Martín, substantially contributed to the STR.

A special thanks goes to the group of professionals that reviewed our initial draft, for their guidance and comments: Eduardo Ayesa, Damian Dominguez, Krist Gernaey, Joe Husband, Paul Lessard, Doug Lumley, Henrik Melcer, Art Umble and Zhiguo Yuan.

We would also like to thank the DOUT workshop contributors for providing additional insights from the wider water resources field:

Charles Bott
David Clark
Tony Flameling
Tony Jakeman
Mark van Loosdrecht
Holger Maier
Libor Novak
Luk Peeters
Antonio Manuel Moreno Ródenas
Simon Tait
Franz Tscheikner-Gratl
Christian Urich

Ludiwine Clouzot acted as our special editor and helped to form the current version of the STR.

We would like to thank the reviewers of the final version of the STR: John B. Copp, Jeffrey Weiss and Erik U. Lindblom. Their edits and suggestions were invaluable and helped us improve the quality of the STR.

Finally, we would like to extend special recognition to the people that inspired us to embark on this journey:

Bruce Beck
Willi Gujer
Paul Harremoës
Peter Reichert
Stefano Saltelli

Introduction to the Scientific and Technical Report

MOTIVATION AND PROBLEM STATEMENT

Over the past 30 years, mathematical models (usually included in simulators) have been displacing the use of heuristic-based ('rules-of-thumb') guidelines for designing water resources recovery facilities (WRRF). Models unify our knowledge about the treatment processes into a single package that is capable of generating comprehensive portraits of how a design will perform. In addition to their value in WRRF design, these models are increasingly being used to evaluate the effect of proposed modifications to plant operations and control, aimed at improving plant performance. Key reasons for the transition to model-aided design and operations include:

- Models allow for more realistic representation of the complexities introduced by hydraulic regime, reactor configuration and variation in operating modes.
- Models can simulate plant performance under dynamic conditions.
- Models enable the designer to analyse and isolate the impact that individual unit operations have on the performance of the treatment train as a whole.
- Models facilitate the examination of effluent quality sensitivity to specific design assumptions.
- Models allow the designer to efficiently screen alternative designs for those that best meet specific environmental goals such as energy efficiency or minimization of greenhouse emissions.
- Models streamline performance comparison of alternative plant designs by facilitating direct comparisons.
- Models can simulate effluent quality response to transient conditions such as wet weather induced influent loadings and operating strategy.
- Models address the growing consensus amongst wastewater professionals that the quality of performance prediction is a critical component of design and operation.
- Wastewater simulation software is a knowledge capture/communication tool that is constantly being updated to simulate new treatment technologies as they gain acceptance and to improve the simulation of existing processes as their behaviour becomes better understood.

A goal of any treatment plant design project is to provide a facility that can be operated reliably to meet specific treatment objectives at minimal cost. Many alternative designs with varying cost structures, performance and risk profiles might be considered to meet the defined objectives. The designer is tasked with the responsibility of finding an acceptable balance between cost, risk and benefit.

Identifying this balance is hampered by the lack of available protocols for explicit risk and benefit assessment. Traditional design methodologies are based upon guidelines and heuristics that have survived the test of time. However, in their application, the complexities of the treatment process are simplified. For example, the variability in plant flow and influent wastewater load is typically addressed through the use of peaking factors. Uncertainty in the coefficients that determine process efficiency is accounted for through the application of safety factors. Judicious choice of these factors provides for a margin of safety that is supposed to ensure adequate performance.

Process model-based design, in addition to the benefits listed earlier, allows the design engineer to incorporate much more information into the design process and in turn to support a more informative assessment of risk and benefits. The reality though is that under current practice, when engineers are interested in evaluating the robustness of a design, they will often overlay a safety factor approach onto the simulation results to accomplish this. In the absence of a prescribed procedure, each engineer will do this in a way that reflects his/her own experiences and prejudices, resulting in some level of arbitrariness.

This need not be. The power and sophistication of existing treatment plant simulators, combined with the wider availability of real-time data and advancements in statistical and data analysis methods, creates opportunities for quantifying treatment plant performance under a wide range of operating conditions. With properly defined protocols, performance profiles can be generated that enable formulation of probabilistic statements (likelihood) regarding various types of failure. Risk/benefit/cost analyses of multiple design alternatives to support identification of the optimal design can be generated. This can be done with a high level of transparency so that each stakeholder can be better informed of the trade-offs they are asked to accept. This is the long-term goal whose exploration is being initiated with this Scientific and Technical Report (STR).

The primary focus of the STR is to develop a comprehensive, workable, and well documented framework for addressing uncertainty and integrating it into WRRF design and operations optimizations. This includes defining what is meant by uncertainty, identifying where uncertainty arises in a project, how uncertainty fits into predicting the long-term performance of a design, how uncertainty influences the attitudes and thus the decision-making process of various stakeholders, methods that are currently available for addressing uncertainty, and methods that are needed but have yet to be developed.

This STR is envisioned as a reference for utilities, regulators and consultants dealing with uncertainty, opportunity and risk in wastewater treatment. The technical details covered in the STR are fleshed out within a comprehensive and holistic framework. This holistic view extends the discussion beyond those uncertainties directly associated with the application of treatment plant models into other areas that influence the final design. This is done in recognition that the chosen design is shaped by inputs from many different stakeholders, and in acknowledgement that uncertainty arises at many stages of project development and execution. These non-model associated uncertainties are important components of the overall uncertainty that influence project risk.

To clarify this last point, consider that the stakeholders in a project might include the public in general, interest or advocacy groups, facility owners, facility operators, facility users, regulators, planners, engineers, designers and contractors. Each comes to the table with different concepts of project objectives and different perceptions and appetites for risk. To illustrate the need for a holistic approach and the complexities of risk in infrastructure projects, consider the example in Box 0.1.

BOX 0.1 UNCERTAINTY – THE BIGGER PICTURE

An engineer is designing a facility that must meet a defined set of effluent limit guidelines. She may receive certain prescriptive criteria for the design from others. She can apply one of the available treatment plant simulators that will enable her to determine the critical aspects of the configuration and sizing of the treatment tanks. She works up a design and then by doing some sensitivity analysis, determines an envelope of conditions under which the design is expected to meet treatment objectives. In doing the sensitivity analysis, she might incorporate some knowledge she has pertaining to statistical uncertainty in some of the model parameters. She might then do some statistical analysis to determine the probability that conditions outside her envelope will be experienced. Based on the findings, she might develop various iterations on the design until she finds a suitable risk profile. The magnitude of that risk is a function of the variability in the key constituents in the wastewater and the uncertainty of various stoichiometric and kinetic parameters in the simulation software being used.

Now the engineer might have received information for the design basis from a planner. The planner may have focused on current and future land use to make forecasts of flows to the plant. He may have decided that the facility should be designed to handle the flows expected 40 years into the future, at which time he expects the catchment to reach maximum flow. To reach this conclusion, the planner may be applying models that are specific to his discipline. He also faces a different set of uncertainties which also contribute to the risk of the project. Whatever the planner determines may simply end up as a specification to which the design engineer must respond, but without any explanation of the attached risks and uncertainties. As a result, embedded into the design are risks unknown to the engineer.

The regulator is charged with setting effluent limitations. In setting limitations, he is guided by the beneficial uses designated for the receiving stream, the water quality objectives necessary to protect those uses and the waste load allocations that follow from those objectives. The regulatory authority might have its own set of models to consult when considering this problem. And these models come with their own unique sources of uncertainty. Then there is the possibility that in the future, the public demands a change in the beneficial uses, or perhaps a future ecological study determines that the assimilative capacity of the receiving is less than originally thought. This might result in a reduction of a waste load allocation with a concomitant lowering of the effluent limitations. How does one consider this regulator risk?

Finally, there are risks that arise out of the contract delivery methods (e.g., has the owner bid the design and construction phases separately or as a package?). Contract delivery methods allocate project risks in different ways and this will have different impacts at various stages of project development.

STRUCTURE OF THE REPORT

This STR is divided into four sections as shown in Figure 0.1 below. Section I, 'System understanding', opens with a general discussion of risk, variability and uncertainty, and identifies how they may influence decisions made at various stages of a project (Chapter 1). This section continues with an assessment of how uncertainty is currently handled in practice (Chapter 2). Reading through Chapter 2 – Current practice, the reader should become aware of the fact that the selection of safety factors and conservative design flow and load values are the most prevalent methods used by engineers currently to account for uncertainty and variability. The section concludes with the benefits of incorporating

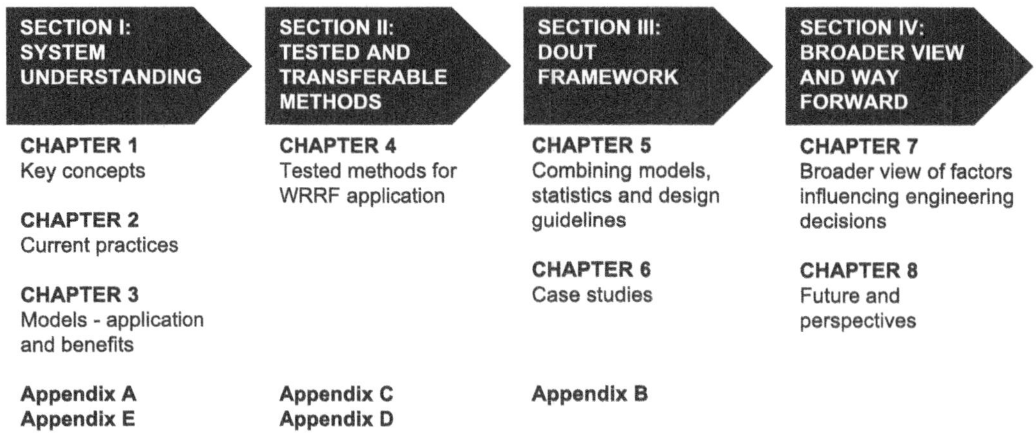

Figure 0.1 The content flow of the STR.

uncertainty analysis in plant design through the use of simulators (Chapter 3). After reading Chapter 3 the reader will have learned about the major sources of uncertainty and variability, how they can be classified in a modelling framework, how practical it is to separate variability from uncertainty and the design parameters that are not amenable to this separation.

Supporting Section I are Appendices A and E. Appendix A includes additional terms and definitions relevant to uncertainty. Appendix E includes examples of engineering practices across selected parts of the world.

Section II, 'Tested and transferable methods', focuses on available methods that allow professionals to manage and evaluate ***quantifiable*** uncertainty in explicit ways. It introduces the reader to concepts and methods which are found in the literature and assesses their feasibility for widespread use in wastewater engineering (Chapter 4). A comprehensive up-to-date literature list has been included in Appendix C. Methods developed in other fields and an assessment of their potential for transfer to the wastewater sector can be found in Appendix D.

Section III, 'DOUT framework', presents a proposed methodology for combining models, statistics and design guidelines for plant design (Chapter 5). The methodology is applied to the case studies presented in Chapter 6. Chapters 5 and 6 focus on two types of uncertainties: quantifiable and scenario uncertainties. Details on the theory behind the methods described as well as further reading materials, can be found in Appendix B.

The final section, 'Broader view and way forward', presents a broader view of the factors influencing engineering decisions (Chapter 7). Chapter 7 discusses the relevance of the contractual environment, the role of the stakeholders and the type of project, and how these play a far greater role in shaping the final outcome of an infrastructure project than is widely acknowledged. Chapter 8 examines possible future ways of dealing with uncertainty and exposes existing challenges, as well as methods available that can already be used by the profession to deal with issues of variability and uncertainty.

Chapter 1

Key concepts of the STR

1.1 INTRODUCTION

This chapter introduces the key concepts of uncertainty analysis which are discussed in this Scientific and Technical Report (STR). Understanding these concepts is a necessary first step in the pursuit of the goal of integrating uncertainty into model-based assessments of water resource recovery facilities (WRRF) design and operations for the purpose of quantifying risk.

The chapter opens with the definition of risk and uncertainty. Uncertainty is a particularly problematic term as it is often used interchangeably with risk, reliability and other similar terms. In addition, these terms may have other meanings in uncertainty evaluations conducted in other disciplines (additional definitions of concepts and terms relevant to the topics covered in this STR can be found in Appendix A). This section also includes how uncertainty has been classified by the scientific community and makes an important distinction between uncertainty and variability.

The chapter closes with a summary of the way uncertainty and variability can be evaluated with the use of models and statistical methods.

1.2 RISK

Risk can be defined as the exposure to events that if realised, result in some sort of loss. Identifying and quantifying – when possible – potential risks is a starting point for risk assessment. Risk, from a traditional engineering perspective, can be quantified as the probability of a specific failure occurring multiplied by the cost of the resulting damage. Therefore, risk quantification has two components: assessment of the probability that the risk will actually manifest, and determination of the associated cost. Certain risks are more amenable to quantification than others. Conversely, other risks can only be assessed qualitatively, for example, by stating that the probability is likely of unlikely, or that the cost of failure is low, medium or high.

Editor(s): Evangelia Belia, Lorenzo Benedetti, Bruce Johnson, Sudhir Murthy, Marc Neumann, Peter Vanrolleghem and Stefan Weijers
doi: 10.2166/9781780401034_0001

Major types of risk in wastewater treatment projects include (Talebizadeh, 2015):

- Non-compliance;
- Loss of reputation;
- Financial loss;
- Not winning a contract or contract annulment.

Such events are called hazards and their expected frequency of occurrence is usually quantified by a probability. Probabilities describe the expected likelihood of occurrence of an event. It answers the question: is it 'likely' or 'unlikely' that the event will happen? Risk is then calculated as the product of the probability of failure and the cost of failure.

Uncertainty assessment and propagation are the methods with which we quantify probabilities and thus, risk.

BOX 1.1 RISK

Risk = [Probability of failure] * [Cost of failure]

To quantify risk the probability of a hazard must be calculated.

This is achieved by assessing uncertainty.

1.3 UNCERTAINTY

Uncertainty can be defined as the degree of inability to determine or predict the exact behaviour of a system or process both now and in the future.

1.3.1 Classification of uncertainty

Uncertainty arises at many points in engineering projects. Although, this has long been recognised, development of a framework for incorporating uncertainty analysis into model-based decision support in wastewater treatment has lagged.

Researchers have classified uncertainty into categories depending on the methods and tools used to quantify or characterise it, in order to provide a common ground for communication between project participants (Refsgaard *et al.*, 2007; Walker *et al.*, 2003). A widely used approach defines three dimensions/categories of uncertainty: nature, location and level. These dimensions are discussed below in greater detail.

1.3.1.1 Nature of uncertainty

The nature of uncertainty refers to whether the uncertainty can be reduced with measurements or further research (e.g., due to experimental uncertainty in the determination of kinetic parameters) or whether it is due to the inherent variability of a system and cannot be reduced with any further research (e.g., frequency of observed events such as heavy rainfall or toxic spills) (see Section 1.3.2).

1.3.1.2 Level of uncertainty

The level of uncertainty is an expression of the scale of uncertainty associated with an identified risk. Based on Walker *et al.* (2003), the Task Group settled on four levels of uncertainty that define a spectrum ranging from complete determinism to indeterminacy (Figure 1.1):

Quantifiable uncertainty can be quantified and described with statistical methods. The random error in a measurement by a sensor, or in the triplicate analysis of a COD (chemical oxygen demand) sample are two examples of quantifiable uncertainty. Quantifiable uncertainty would also include the error in estimating a population mean from a set of samples.

Scenario uncertainty is uncertainty associated with the use of scenarios to examine possible outcomes that may develop in the future. Scenarios do not forecast what will happen in the future. Instead, they assess what might happen. Realistic assumptions about relationships and/or driving forces within the model can be established. It is not possible, however, to derive the probabilities of the scenarios taking place. Scenario uncertainty can be presented as the range of discrete outcomes from a scenario analysis.

Recognised ignorance is the state where the existence of uncertainty is recognised, but the uncertainty does not lend itself to quantification, nor to study by means of scenario analysis. In this situation, the mechanisms and functional relationships of the phenomena impacted by uncertainty are too poorly understood to enable any useful analysis beyond recognising that there is uncertainty, but it cannot be characterised.

Total ignorance is defined as the state where those involved are not aware of uncertainty. It is unknown what is unknown.

Figure 1.1 depicts these four levels of uncertainty lying between determinism and indeterminism.

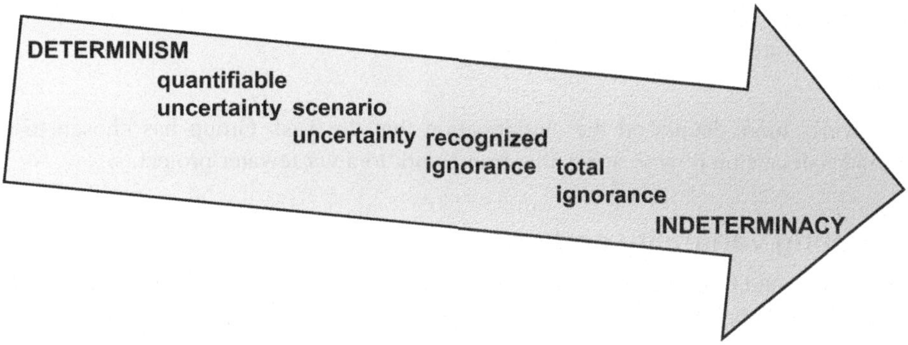

Figure 1.1 Level of uncertainty (after Walker *et al.*, 2003).

1.3.1.3 Location of uncertainty

The location of uncertainty refers to the instance where uncertainty manifests itself in the modelling process. Walker *et al.* (2003) suggested five generic locations: context uncertainty, model uncertainty, input uncertainty, parameter uncertainty and model output uncertainty. Walker's conceptualisation of these uncertainties is elaborated upon below.

- **Context uncertainty**: Context refers to the economic, political, social and technical conditions and circumstances that influence the model boundaries and frame the issues that the model is to address. Context uncertainty also relates to the suitability of a model for its intended purpose.
- **Model uncertainty**: All models involve simplifications of the system under study. Model structure uncertainty refers to uncertainty as to whether the model is an adequate representation of the real system it represents. In addition, errors can arise when implementing a model into a simulator. This is associated with the translation of the model into a program code and its execution on a computer and also includes uncertainty due to software errors.

- **Input uncertainty**: Input uncertainty is comprised of two sub-categories, system data uncertainty and external driving force uncertainty. Data uncertainty include uncertainty in, for example, the flow and concentrations to be input to a model for the purpose of projecting plant behaviour under some future condition. External driving force uncertainty relates to uncertainty associated with changes in conditions that are outside the model boundaries. For example, land-use policies in the catchment to a treatment plant might change and open the catchment to more rapid development. The knock-on effects from this would result in a change in flow and wastewater characteristics for the plant.
- **Parameter uncertainty**: Treatment plant models include many kinetic and stoichiometric parameters. The values of many of these parameters are known only approximately. Parameter uncertainty is associated with the lack of knowledge regarding the true value of these parameters as well as the different techniques used for the selection of parameter values during model calibration.
- **Model output uncertainty**: Model output uncertainty is the accumulated uncertainty caused by all the uncertainties in all the above locations as propagated through the model.

The Task Group chose to modify the Walker *et al.* (2003) framework. Context uncertainty is placed outside of the scope of uncertainties that are to be considered in this STR for model-based decision making. 'Source' of uncertainty is used instead of 'location' of uncertainty. The Task Group has chosen to organise the sources of uncertainty as follows:

- Inputs (includes experimental error);
- Model structure;
- Numerics (software implementation issues);
- Model output.

Section 1.4 provides more details on the classification that the Task Group has chosen to implement. Chapter 5 provides details on how to apply this framework to a wastewater project.

1.3.2 Separating variability and uncertainty

As discussed by Kelly and Campbell (2000), the EPA risk guidance and policy document (US EPA, 1997) and the report by the National Academy of Sciences titled Science and Judgment in Risk Assessment (NRC, 1994) call for separating variability and uncertainty in risk assessments.

There is an important difference between variability and uncertainty (and which quantities should be considered variable, uncertain or both). Although both can be described mathematically in the same way, for example by using density functions (Figure 1.2), they are very different in nature (Table 1.1). The Task Group has selected the definitions included in Box 1.2, in order to clarify the confusion often seen in the literature.

In this STR, uncertainty which is classified as irreducible (also aleatoric uncertainty), is designated as variability whereas uncertainty classified as reducible (also epistemic uncertainty), will be referred to simply as uncertainty.

1.3.3 Sources of variability and uncertainty

The WERF (Water Environment Research Foundation) study: 'Evaluating the Performance of Nutrient Removal Treatment Processes' (Bott & Parker, 2011) highlighted that plant performance variability depends on site-specific conditions: 'Local conditions impact the performance achieved on average and in terms of statistical variability. These factors include process design, climate impacts, wet weather flow influences, attributes of the service area, variation in influent flows and loadings, presence or absence of industrial contributions, whether solids processing is accomplished on the same site, sustained or interrupted supplies of chemicals, construction impacts, mechanical failures, the difficulty in operating the process, the ability to automate the controls of a process, the closeness of operation to design flows and loadings and others'.

The study examined plant data and identified specific factors that can be classified as external and linked to the 'environment' or internal and linked to the 'system'. The environment was detailed as the (past and future) inputs to the treatment plant as well as the responses of the receiving water body to the outputs of the treatment plant. The system was the WRRF.

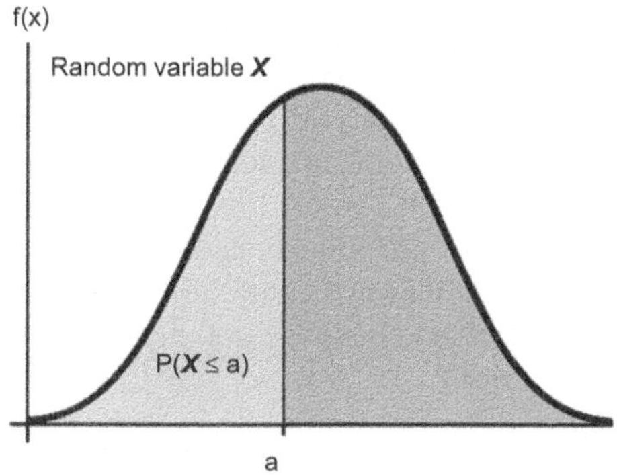

Figure 1.2 A density function can be used to represent either uncertainty or variability.

Table 1.1 Distinction between variability and uncertainty.

	Uncertainty	Variability
Origin	Lack of knowledge	'Real spread' of values (in time or space)
Reducibility	Partly reducible by further investigation	Irreducible
Representation	Probability distributions, density function	Time series, frequency distributions, density function
Example	Triplicate analysis of a COD sample, substrate hydrolysis rate	Influent COD (e.g., daily loads of one year)

BOX 1.2 VARIABILITY VS. UNCERTAINTY

Variability: is defined as the 'real spread' of values (in time or space) of a well-specified (statistical) population (Example: observed daily average COD load in the influent of a specific treatment plant over 5 years). The spread of these values is not reducible by further knowledge acquisition. Variability is a property of the population, not of the state of knowledge (Kelly & Campbell, 2000).
Uncertainty: results from a lack of knowledge. Parameter uncertainty is the uncertainty about the appropriate values for model parameters (e.g., half-saturation constants, hydrolysis rates). Model structure uncertainty pertains to the adequacy of the model equations and the model resolution in view of the modelling objective. Unlike variability, uncertainty is partly reducible: for example, further measurements or deeper investigations into the relevant processes might increase knowledge.

Examples of sources of variability and uncertainty originating from the environment are:

- Climate effects, energy use (resulting in operating strategy changes) and collection system/sewer characteristics;
- Wastewater characteristics (flows, loads, temperature, alkalinity, pH, fractionation);
- Growth or loss within the collection area (growth rate, changes in inflow and infiltration, changes in industry);
- Discharge permits.

Examples of sources of variability and uncertainty originating from the system are:

- Biological: microbial growth behaviour, especially at low concentrations;
- Physical: effect of unit operations configurations on removal efficiencies, alpha value (in aeration);
- Physical: non-ideal process behaviour (transport phenomena in aeration systems, non-ideal mixing affecting plant performance, approximation of plug flow hydraulics by using continuous stirred tank reactors (CSTRs) in series);
- Physical–chemical: for example, precipitation stoichiometry and kinetics;
- Biological-colloid chemical: effect of load and composition variations/peaks on sludge composition and floc structure and subsequent effect on sludge sedimentation;
- Unexpected control system behaviour;
- Mechanical failure;
- Operational problems.

1.3.4 Uncertainty analysis approaches

The role that a professional plays within a project influences where she will direct her focus on questions involving variability and uncertainty and the approach to uncertainty analysis that she will prefer. A project manager, who will oversee all aspects of project execution, might want to know how uncertainty affects the critical decisions affecting project development. A risk manager may hone in on uncertainties related to the type of contract and its influence on the risk to the various stakeholders. An engineer may break the project down to its various phases and then move to identify the sources of uncertainty within each phase. A scientist might focus on uncertainties in the data and the mathematical model that will inform the facility design. Tables 1.2–1.4 present three possible (not mutually exclusive) ways of addressing uncertainty (i–iii).

(i) Through project phases (Table 1.2);
(ii) Through modelling project steps (Table 1.3);
(iii) Using a systems analysis framework (Table 1.4).

Table 1.2 Examples of sources of uncertainty through infrastructure project phases (type i).

Phase	Example
Regulatory	Future effluent limits
Planning	Design horizon, design load
Preliminary design	Configuration type, critical growth rate, yield
Detailed design	Number of pumps, aerator layout
Construction	Unexpected geotechnical issues
Commissioning	Stability of processes
Operations	Toxic spills, foaming and bulking, sludge settling

Table 1.3 Examples of sources of uncertainty across the steps of a modelling project (type ii).

Phase	Example
Project definition	System boundary, required prediction accuracy
Data collection	Representativeness of historical data
Plant model-setup	Choice of biological model
Calibration/validation	Model parameter values
Simulation	Choice of scenarios, uncertainty propagation settings

Table 1.4 Examples of sources of uncertainty in a systems analysis framework (type iii).

Phase	Example
Aggregation/sampling error	Point measurements of rainfall
Measurement error	Random, systematic, gross errors
Input uncertainty	Catchment behaviour
Parameter uncertainty	Kinetic, mass-transfer related
Model structure	Monod vs. Haldane kinetics
Numerical	Insufficient numeric resolution

Researchers are typically more familiar with type (iii) structuring of uncertainty within a systems analysis framework whereas practitioners will normally be at ease with type (i) structure. Type (ii) can be interpreted as a combination of (i) and (iii) and reflects the decisions typically taken by the modeller (see also Rieger *et al.*, 2013). It is important to note that a modelling project could be implemented for any of the engineering project phases (see Chapter 5).

As the STR has the objective to facilitate the transfer of methods from research to practice, it is helpful to structure the sources of uncertainty within a framework which practitioners can easily relate to. Therefore, the first two approaches are emphasised.

Chapters 2–6 discuss the use of models for the evaluation of variability, quantifiable (statistically) uncertainty and scenario uncertainty.

1.4 INCORPORATING VARIABILITY AND UNCERTAINTY ANALYSIS IN MODELS

Mathematical models (such as the IWA Activated Sludge Models (Henze *et al.*, 2000)) coupled with statistical methods can assist practitioners in assessing variability and quantifiable uncertainty during plant design and operation. This section focuses specifically on uncertainties associated with the application of process models. Models can be used in any one of the project phases discussed in Table 1.2.

1.4.1 Variability and uncertainty in model steps

Rieger *et al.* (2013) in developing standards for Good Modelling Practice, defined a five-step model development protocol. Uncertainties that arise at each step are shown in Figure 1.3 and discussed in the following sections.

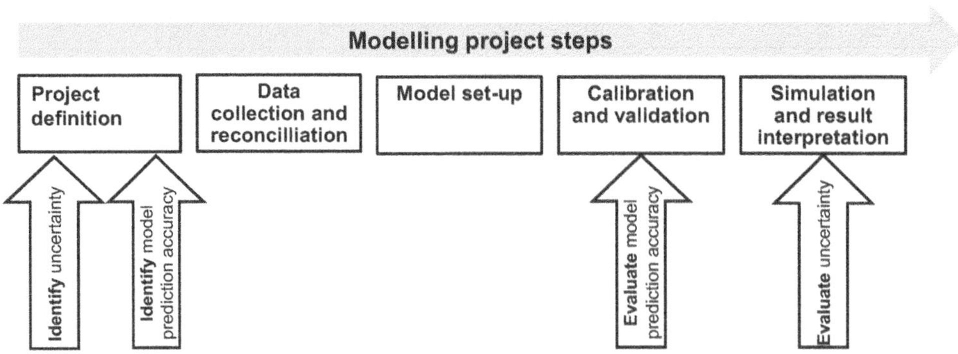

Figure 1.3 Modelling project steps and sources of uncertainty.

Project definition involves identifying the goals of the modelling project and setting the criteria for predictive accuracy. Based on the engineering project phase the pertinent sources of uncertainty will be identified in this stage.

Uncertainties associated with the project definition will differ depending on the project phase for which the modelling exercise is conducted (see Table 1.2 for project phase). For example, greater model accuracy will be demanded if the model is built to support detailed design rather than preliminary design. Also, at later project phases, more information will have been acquired and this will tend to reduce uncertainty.

Data collection and reconciliation is performed to improve data quality by removing noise and other artefacts in the data. This is necessary as raw data collected for the model will normally contain errors, outliers and gaps. Data gaps may be filled in using interpolation or other methods.

Uncertainties associated with data collection and reconciliation include questions as to whether the data are suitable for the intended model application. For example, the data might contain so many gaps that filling those gaps results in an undesirable skew.

Plant model set-up, also known as structure identification, refers to the identification/selection of an appropriate model structure. The definition of model structure includes reactor design (e.g., fixed-bed reactor vs. suspended growth) and reactor configurations (e.g., anoxic and aerobic zones, plug flow vs. CSTR) and biokinetic model selection (e.g., ASM or Monod/inhibition terms).

Uncertainty about the exact reactor and plant configurations (e.g., dimensions), and the true reaction mechanisms, results in a model structure which is subject to error, known as model structure uncertainty.

Calibration or parameter optimisation of a model consists of the adjustment of its numeric parameters, for example, kinetic parameters, to fit observed data. The purpose of calibration is to obtain a model that better reflects observations made under a specific set of conditions (operation).

Validation of a model consists of the evaluation of the model performance on observed data by comparing the simulation results of a calibrated model to an independent set of observations.

Uncertainty in the data (e.g., measurement errors) can result in model parameter values that are different from their true values.

Simulation and result interpretation relates to the application of a model to evaluate process performance (e.g., effluent ammonia concentrations are simulated). Once a model is finalised, it is often used to generate predictions of plant performance (based on currently available information) under some future condition to assess whether design and operation objectives will be met.

All the uncertainties encountered during model development will propagate through the model leading to model output uncertainty. As defined previously, model output uncertainty is the difference between the predicted values and the response of the real plant when operated under the conditions reflected in the model inputs. The more closely the model represents the real plant process and the more representative the input data, the better the model predictions and the less uncertainty.

The steps described above apply to projects where models are used for existing plants, where influent and plant data are available. In green field design situations where data are often not available, the data collection and reconciliation step and the calibration and validation steps are omitted. Data from nearby plants or similar catchments and default model parameter values can be used. However, regardless of the application, the uncertainties described above still exist. Chapter 5 discusses in more detail the uncertainty-related tasks that need to be considered at each stage of a wastewater treatment modelling project.

1.4.2 Sources of variability and uncertainty in models

1.4.2.1 Model input variability

Input variability results from the variable pattern exhibited by an input variable to the model (e.g., variable rainfall in a catchment). This input variability when propagated through the model, will cause variability in predicted plant performance (Figure 1.4).

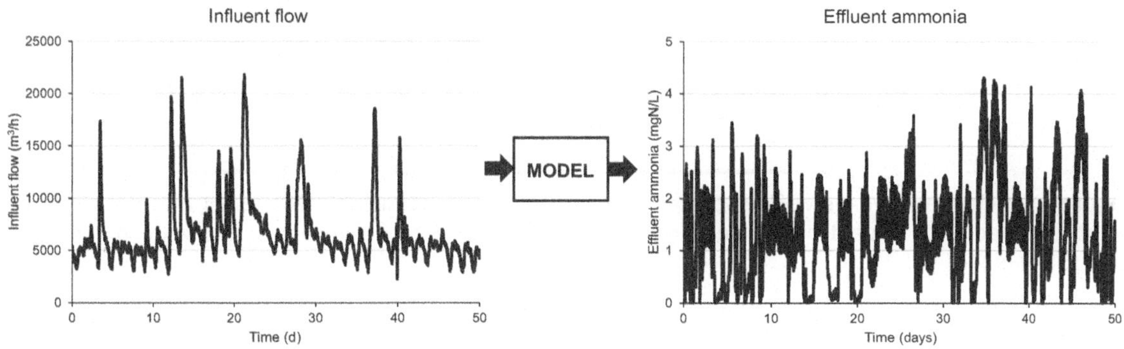

Figure 1.4 Impact of influent variability on effluent ammonia.

1.4.2.2 Model input uncertainty

Input uncertainty is a result of, for example, observation or analytical error and results from the variability of repeated measurements. Another source of input uncertainty is a result of filling in data gaps by interpolation or other methods. Data filling introduces varying errors which propagate through the model and affect the uncertainty of the model output.

1.4.2.3 Model structure uncertainty

Model structure uncertainty can be defined as uncertainty in model predictions originating from assumptions and simplifications made in the structure of the mathematical model. A mathematical model is always a simplified representation of reality. This leads to some degree of uncertainty in the model output predictions, originating from the process detail or rigour that is missing in the model. Uncertainties with the model structure are often associated but not limited to the following model selections:

- Influent fractionation model;
- Biological and chemical model;

- Hydraulic model;
- Aeration system model;
- Clarifier model;
- Models of sensors, actuators and equipment in plant operations;
- Interfaces between models.

Figure 1.5 includes a schematic depiction of the sources of model structure uncertainty for an example activated sludge plant that includes aeration tanks and final clarifiers. The sources have been sub-divided based on their spatial description (hydraulics) and the conversion process description (influent, biological or settling model).

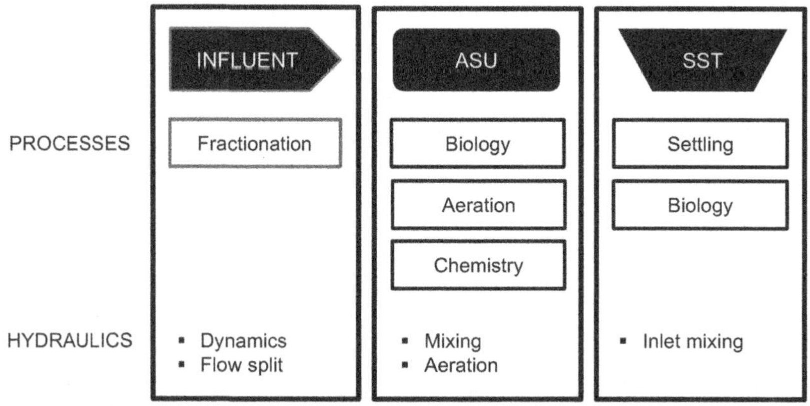

Figure 1.5 Schematic overview of the sources of uncertainty for an example WRRF model. The blocs indicate sources related to the influent model, the activated sludge unit (ASU) model and the secondary sedimentation tank (SST) model.

It should be noted that in whole plant models plant configuration simplifications also introduce model structure uncertainty.

1.4.2.4 Model parameter uncertainty

Treatment plant models contain many parameters that must be assigned values prior to model execution. Lack of knowledge regarding the best parameter value for a given system is an example of model parameter uncertainty. This is especially true for new processes that have not been properly characterised through multiple model development efforts. Examples of parameters without scientific consensus include parameters in models for nitrous oxide production, de-ammonification and the use of alternate substrates for denitrification.

Parameters that introduce uncertainty in biological process models are typically divided into kinetic (rates) and stoichiometric (coefficients). An example of a stoichiometric parameter is yield while, examples of kinetic parameters include bacterial growth, decay and hydrolysis rates as well as their associated half saturation constants. Rate parameters become important when the associated substrate or electron acceptor becomes 'growth limiting' and half saturation coefficients are critical for process sizing. Stoichiometric parameters are important for determining how a mass of reactant is distributed amongst one or more products for any given transformation. A very important set of parameters that introduce uncertainty are the parameters and ratios included in the influent fractionation models.

Some parameters are known with greater confidence (lower uncertainty) than others. For example, the yield of ordinary heterotrophic bacteria (OHO) on readily biodegradable (fermentable) substrate (S_B) growing aerobically is well characterised. A value in the range 0.60–0.67 g of OHO COD per gram of S_B consumed has been found to work well in modelling projects. However, uncertainty in the value of yield might be important when alternative substrates such as methanol, ethanol or glycerol are used for denitrification as these have not been as widely researched. As yield relates the consumption of carbon to electron acceptor consumption any uncertainty in the alternative substrate yield estimate could significantly affect the modelled economics of using one of these substrates for denitrification.

Whereas the standard yield is well established, the values of other standard parameters are not well known at all. For example, when modelling hydrolysis, the conversion of particulate biodegradable COD to S_B, plays an important role in the prediction of the process behaviour for systems that have a very short SRT or for processes that largely metabolise particulate substrates (aerobic and anaerobic digestion). Because conditions are never constant developing experiments to accurately estimate the hydrolysis rate under all possible conditions is essentially impossible which can introduce considerable uncertainty in the model predictions when this process is critical to the model output.

This section has illustrated the variability in parameter knowledge using a couple of simple examples, but the reader is reminded that models have hundreds of parameters and each has a different level of uncertainty and each of these parameters has a different impact on the model predictions. This realisation leads to the conclusion that it is important to carefully consider the level of uncertainty in each parameter and ultimately understand the impact that that uncertainty may or may not have on the model output.

1.4.2.5 Numerical uncertainty

Process model simulation is carried out in a number of consecutive steps. First, the real process is described by a mathematical model (a process model using ordinary or partial differential equation (ODE or PDE)). This is then approximated by a simulation model (numerical method) to be implemented in a computer (Bürger *et al.*, 2011). In a third step, the simulation model needs to be implemented in a software platform. The lumped uncertainty coming from approximating the mathematical model by a simulation model and its implementation is called numerical uncertainty. Each of the steps includes errors and approximations resulting in numerical uncertainty. As discussed by Claeys *et al.* (2010), no automatic tool exists to quantify this uncertainty.

The main sources of numerical uncertainty (Figure 1.6) can be classified as either numerical, implementation uncertainty, coding uncertainty, solver suitability, solver coding uncertainty and machine uncertainty (based on Claeys *et al.*, 2010).

The only source of uncertainty a model user can influence directly is the choice of solver and its accuracy settings (solver suitability). Claeys *et al.* (2010) show that this source of numerical uncertainty (caused by

Numerical uncertainty					
Source of uncertainty	Solver coding uncertainty	Machine uncertainty	Numerical implementation uncertainty	Coding uncertainty	Solver suitability
Type of modeller	Software developer	Software developer	Model implemeter	Model implemeter	Model implemeter Model user

Figure 1.6 Overview of the different sources of numerical uncertainty and modeller type that can influence this uncertainty (based on Claeys *et al.*, 2010).

the incorrect application of the solvers) can be more important than parameter uncertainty. As there are no analytical solutions for most (if not all) of the ordinary differential equations (ODEs) and partial differential equations (PDEs) used in wastewater treatment modelling, numerical methods are used to approximate the solution for these models.

The solver coding uncertainty and machine uncertainty can only be prevented by the software developers. Since the model is implemented in a computer and computers have a finite precision floating point arithmetic, any computation involves rounding errors. So, machine uncertainty is caused by the rounding errors and can become very important in iterative processes such as numerical integration.

Numerical implementation uncertainty and coding uncertainty are sources of numerical uncertainty that are introduced when a mathematical model is implemented in a software platform (Hauduc *et al.*, 2010). This implementation is executed in two steps, the translation of the model into a simulation model and the actual coding work. In both steps, conceptual and technical errors, leading to uncertainty, can be introduced.

1.4.2.6 Model output uncertainty

As discussed earlier, model output uncertainty is the accumulated uncertainty caused by all the uncertainties in all the above locations as propagated through the model. Model output uncertainty can be defined as epistemic (reducible) uncertainty and it relates to the differences between the true values of the output quantities and the values predicted by the model. Quantification of model output uncertainty serves as qualification and acceptance of the models used; whose outputs inform a model-based decision-making process.

1.4.3 Evaluation methods

Monte Carlo simulation, expert knowledge with fuzzy logic and optimisation are among the most widely used methods for exploring the combined effects of how the various sources of uncertainty propagate through the model and affect model output. The project phases where these methods can be used are listed in Table 1.5. The methods are critically reviewed in Chapter 4 and additional technical details have been included in Appendix B.

Table 1.5 Methods used for model uncertainty analysis.

Phase	The Most Used Methods	Main Applications
Planning	Monte Carlo Expert knowledge and fuzzy logic Scenario analysis	Output requirements Technology selection Scenario analysis
Preliminary design	Monte Carlo and mixed optimisation methods	Performance evaluation Plant dimensioning Control system selection
Detailed design	Monte Carlo combined with: (i) CFD, (ii) optimisation	Exact dimensions Control system design
Operations	Monte Carlo	Process analysis Optimisation Control

Monte Carlo analysis is the dominant method used to evaluate quantifiable (epistemic) uncertainty and can be applied across all stages of design. In the early stages of design, when there are many degrees of freedom and more uncertainty, Monte Carlo and scenario analysis are the preferred techniques. As the design process progresses (and the degrees of freedom are reduced), Monte Carlo is still the dominant technique, but it is sometimes coupled with mixed optimisation-based techniques (GA, NLP, Pareto frontiers).

Expert knowledge and fuzzy logic have also been used, but mostly for technology selection and scenario analysis.

Optimisation methods are typically used in preliminary and detailed design to develop exact dimensions or during operational assessment to develop control settings and process optimisation.

Chapter 4 reviews the uncertainty analysis methods described in the literature in the field of wastewater. Appendix D discusses pertinent methods from other fields.

1.5 SUMMARY

This chapter presented the important concepts necessary to acquire baseline understanding of the classification of uncertainty and how it relates to process modelling. The key points from this chapter can be summarised as follows:

Uncertainty mainly stems from imperfect or unknown information (lack of knowledge) and can be reduced by more research. Uncertainty can be classified as quantifiable or unquantifiable based on whether statistical methods and models can be used to evaluate it.

In this STR, uncertainty which is classified as irreducible (also aleatoric uncertainty), is designated as variability. Variability is the 'real spread' of values (in time or space) of a measurable quantity.

In this STR, uncertainty classified as reducible (also epistemic uncertainty), will be referred to simply as uncertainty.

Quantifiable (statistical) uncertainty in reference to the value of a parameter or quantity can be described with a probability density function.

REFERENCES

Bott C. B. and Parker D. S. (2011). Nutrient Management Volume II: Removal Technology Performance & Reliability, WERF Research Project under Nutrient Challenge, WERF Report NUTR1R06 k, Water Environment Research Foundation, Alexandria, VA, USA.

Bürger R., Diehl S. and Nopens I. (2011). A consistent modelling methodology for secondary settling tanks in wastewater treatment. *Water Research*, **45**, 2247–2260.

Claeys P., van Griensven A., Benedetti L., De Baets B. and Vanrolleghem P. A. (2010). Conceptual thinking on numerical solver selection and uncertainty terminology. *Journal of Hydroinformatics*, **12**(3), 241–250.

Hauduc H., Rieger L., Takács I., Héduit A., Vanrolleghem P. A. and Gillot S. (2010). A systematic approach for model verification: application on seven published activated sludge models. *Water Science and Technology*, **61**(4), 825–839.

Henze M., Gujer W., Mino T. and van Loosdrecht M. (2000). Activated Sludge Models ASM-1, ASM-2, ASM-2d and ASM-3. Scientific and Technical Report No. 9, International Water Association (IWA), London, UK.

Kaplan S. and Garrick B. J. (1981). On the ouantitative definition of risk analysis. *Risk Analysis*, **1**(1), https://doi.org/10.1111/j.1539-6924.1981.tb01350.x

Kelly E. J. and Campbell K. (2000). Separating variability and uncertainty in environmental risk assessment – making choices. *Human and Ecological Risk Assessment: An International Journal*, **6**(1), 1–13.

NRC (National Research Council) (1994). Science and Judgment in Risk Assessment. Committee on Risk Assessment and Hazardous Air Pollutants. National Academy Press, Washington, DC.

Refsgaard J., Van der Sluijs J., Hojberg A. and Vanrolleghem P. (2007). Uncertainty in the environmental modelling process – a framework and guidance. *Environmental Modelling & Software*, **22**(11), 1543–1556.

Rieger L., Gillot S., Langergraber G., Ohtsuki T., Shaw A., Takacs I. and Winkler S. (2013). Guidelines for Using Activated Sludge Models, Scientific and Technical Report No. 22, International Water Association (IWA), London, UK.

Talebizadeh M. (2015). Probabilistic Design of Wastewater Treatment Plants. PhD thesis, Département de génie civil et de génie des eaux, Université Laval, Québec, QC, Canada.

USEPA (U.S. Environmental Protection Agency) (1997). Policy for Use of Probabilistic Analysis in Risk Assessment at the U.S. EPA and Guiding Principles for Monte Carlo Analysis. Office of Research and Development. EPA/630/R-97/001. https://www.epa.gov/risk/policy-use-probabilistic-analysis-risk-assessment-epa and https://www.epa.gov/risk/guiding-principles-monte-carlo-analysis (sites accessed 2021).

Walker W. E., Harremoes P., Rotmans J., van der Sluijs J. P., van Asselt M. B. A., Janssen P. and Krayer von Krauss M. P. (2003). Defining uncertainty: a conceptual basis for uncertainty management in model-based decision support. *Integrated Assessment*, **4**(1), 5–17.

Chapter 2

Uncertainty in wastewater treatment – current practice

2.1 INTRODUCTION

The wastewater treatment industry has evolved towards increasingly sophisticated, capital-intensive engineered systems. Decisions on plant design and operation often depend on the estimation of risk. As discussed in Chapter 1, risk is intrinsically related to uncertainty. By reducing, where possible, uncertainty, the probability of failure can be more accurately assessed and in turn optimal designs can be proposed. To understand risk, we must therefore explore uncertainty.

This chapter covers how risk and uncertainty are currently handled in engineering practice and focuses on the risk of non-compliance. The approaches described in the following sections will be familiar to engineers across the world, however, the chapter focuses on current practice in North America. Examples of design practices in other jurisdictions have been included in Appendix D.

2.2 GENERAL APPROACHES FOR ADDRESSING UNCERTAINTY IN WASTEWATER TREATMENT

2.2.1 Design guidelines

2.2.1.1 Overview

Uncertainty and risk of non-compliance is frequently handled in wastewater treatment practice through the use of design guidelines. Historically, process design criteria have been based on regulatory requirements, industry-accepted design standards or state-specific regulations (industry standards, adapted to specific state conditions with additional requirements). Some examples of these design standards include:

- Theory, Design and Operation of Nutrient Removal Activated Sludge Processes (Ekama *et al.*, 1984);
- Water Environment Federation Manual of Practice 8 (WEF MOP-8, 2017);

Editor(s): Evangelia Belia, Lorenzo Benedetti, Bruce Johnson, Sudhir Murthy, Marc Neumann, Peter Vanrolleghem and Stefan Weijers
doi: 10.2166/9781780401034_0015

- Wastewater Engineering: Treatment and Resource Recovery 5th Edition (Metcalf & Eddy Inc. *et al.*, 2013);
- Great Lakes Upper Mississippi River Board, Recommended Standards for Wastewater Treatment Facilities (Ten State Standards) (GLUMRB, 2014);
- ATV Guidelines (ATV, 2000);
- USEPA Nitrogen Control Manual (USEPA, 1993);
- USEPA Phosphorus Removal Design Manual (USEPA, 1987a);
- Biological Wastewater Treatment (Grady *et al.*, 2011);
- WERF/CRTC Methodologies for Evaluating Secondary Clarifier Performance (Wahlberg, 2004);
- Virginia's Sewage Collection and Treatment Regulations (Virginia DEQ, 2008);
- Biological Nutrient Removal (BNR) Operation in Wastewater Treatment Plants (WEF MOP 29, 2005).

Design guideline documents typically provide design targets such as surface overflow rates for average and peak hour flows. These standards tend to address risk by using relatively conservative design criteria and forcing the designer/engineer to look at multiple scenarios. What these design criteria generally do not do, is address plant-specific conditions or provide methods for determining design flows and loads that are both 'real' and statistically rigorous. Frequently, how the criteria are to be applied is open to interpretation from the designer/engineer and/or regulator. For instance, often these standards do not directly address covariance (correlation) of flows and loads, the interaction between unit processes, or reliability. None of the guidance documents provide specific criteria for managing risk in process design, however using the design standards generally results in a conservative design with relatively low probability of failure.

Because the criteria are frequently open to interpretation, engineers tend to evaluate several scenarios that include combinations of critical design parameters. This approach can result in conservative and expensive designs without necessarily providing a worthwhile benefit (Doby, 2004). Russell (2019) states that most municipal wastewater treatment plants are 30–50% overdesigned based on municipal codes and, after safety factors are used by consultants, can be overdesigned by 100% or more (Box 2.1).

2.2.1.2 Design criteria

In certain jurisdictions, permit writers need to review engineering design reports and contract documents during the permit application process in order to issue construction permits.

For example, in the USA, the listing of specific design criteria for unit processes varies depending on the State. Texas and Virginia, for example, provide design requirements as part of state law for various methods of activated sludge treatment, clarification, and biosolids treatment. The risk for the permit writer is mitigated because the state law mandates criteria that does not leave room for interpretation. Other states, such as Florida, do not list specific design criteria for unit processes in State Code. In this case, the permit writer is dependent on the guidance from other documents such as the Ten States Standards (GLUMRB, 2014) for reviewing design criteria. In Florida, as design criteria are not mandated in the Florida Administrative Code (Fla. Admin. Code, 2013), it is the responsibility of the engineer-of-record signing and sealing the engineering report and/or contract documents to address risk in the design. In this case, risk is shifted from the regulator to the engineer-of record.

2.2.1.3 Safety factors

Historically, safety factors have been the most common approach for mitigating risk for multiple reasons. With early wastewater treatment plants design, safety factors could easily be used to account for a great deal of uncertainty in all of the factors that control process performance.

Lawrence and McCarty (1970) derived a formula for the minimum sludge age to avoid washout. The safety factor was defined as the ratio of the design sludge age to the minimum sludge age (Lawrence, 1971a; Lawrence & McCarty, 1970). German design standards (ATV, 2000) recommend safety factors for determining the aerobic sludge age for nitrification and nitrification/denitrification facilities.

Box 2.1 shows how multiple safety factors may work together to affect the target design concentration.

BOX 2.1 COMPOUNDING OF SAFETY FACTORS

In wastewater treatment infrastructure, safety can be introduced in various ways. A regulator can introduce safety by lowering the discharge permit limit. Similarly, a design engineer can choose to design for an effluent target concentration lower than the permit. Figure 2.1 shows how multiple safety factors may work together to affect the target design concentration. The development of the effluent permit in the regulatory stage includes safety factors to reduce the probability of negatively impacting the ecosystem whereas the WRRF planning and design stages introduce safety to increase the probability of not surpassing this permit. As a result, compounding of safety factors occurs.

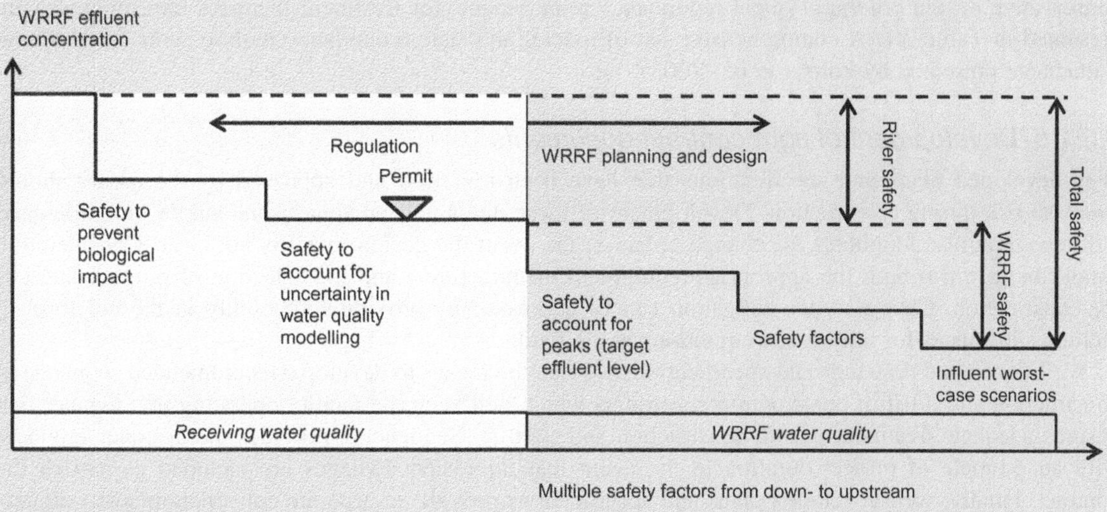

Figure 2.1 Multiple safety factors introduced at different locations reduce the target design effluent concentration.

As our knowledge of the wastewater treatment process has progressed, more sophisticated methods for design (such as modelling) have evolved which give practitioners the means to evaluate reducing safety factors. Models can also be used in combination with safety factors to reduce the number of scenarios that need to be analysed or modelled. For instance, a model can be run for a maximum month condition, with safety factors that are built in to account for daily or diurnal fluctuations in flows and loads.

2.2.1.4 Reliability and redundancy standards

Reliability and redundancy standards are used to reduce the risk of failure due to individual unit processes being out of service either due to mechanical failure or maintenance. Guidelines where available are, by necessity, somewhat vague because they must deal with a wide range of conditions. The real degree of reliability and redundancy is developed by the designer/engineer in conjunction with the owner. These requirements are often difficult to specify as they often rely on the relationships between the equipment at various locations within a facility.

Design guidelines for adding redundant or backup components to a system design have been developed to ensure that critical components retain redundant configurations in the event of failure (Palmer *et al.*, 2003). A definition of redundancy and its links to uncertainty and reliability can be found in Appendix A.

In North America, redundant design practices began with fault-tolerant requirements under the directive of water quality regulatory compliance and safety (USEPA, 2000). The justification for including redundant equipment in treatment facilities began with the need to continuously operate treatment processes during equipment failure and while performing preventative maintenance that required equipment to be taken out of service. This assured that a treatment facility's continuous treatment operations would meet federal and state regulatory requirements that protect the environment and public health and safety.

The U.S. Environmental Protection Agency (EPA) and several states have developed standards for equipment redundancy considerations. These standards tend to address risk by using relatively conservative design criteria. Typical redundancy requirements for treatment facility's unit processes are presented in Table 2.1. A comprehensive list of federal and state redundancy requirements for U.S. and Canada are presented by Palmer *et al.* (2003).

2.2.1.5 Development of tight contract documents

Well-developed plans and specifications that have been reviewed and approved by the owner should decrease risk during construction. Development of these items will prevent errors and/or omissions and will also minimise loopholes for change orders in the event the design intent is not clear. Consideration should be given to both the appropriate equipment manufacturers and the selection of proper materials for construction. Change order mitigation can be addressed by providing flexibility in the bid form to include allowances for unforeseen conditions in the field.

Well-developed drawings and specifications provide the means to develop a recommended sequence of construction and a list of construction constraints which will keep the facility operating in a manner that ensures adequate treatment during construction and start-up. Sequencing provides the designer/engineer with an estimate of project duration in the event that liquidated damages are included as part of the contract. Finally, well-developed plans and specifications provide an accurate construction cost estimate. This estimate will be used by the owner for financing the project.

2.2.1.6 Staffing and monitoring

In certain jurisdictions, under conditions where there may be concern over meeting effluent permit limits, regulators have the flexibility to require additional staffing and/or monitoring known as 'reasonable assurance'. The addition of staff or the requirement for additional monitoring offsets risk and uncertainty that the permit writer may believe is evident during review of the engineering reports and/or contract documents during the permit application process.

Table 2.1 Example of regulatory redundancy/reliability requirements for wastewater treatment facilities.

Unit Process	USEPA (1974)	Great Lakes Upper Mississippi River Board (Ten State Standards) (2014)
Mechanically cleaned screens	One backup screen	Minimum of two screens, with capability to treat design flow with one unit out of service
Pumping systems	One backup pump for each system performing the same function	One backup pump for each system performing the same function
Grit removal	Not specified	Minimum of two units with ability to bypass. No redundant tankage required
Primary sedimentation basins	Must be able to process 50% of plant flow with one (largest) unit out of service	Requires multiple units capable of independent operation for plants with flows higher than 0.1 MGD, but no redundant unit required
Secondary sedimentation basins	Must have 75% of rated capacity with one (largest) unit of service	Requires multiple units capable of independent operation for plants with flows higher than 0.1 MGD, but no redundant unit required
Activated sludge – Aeration basins	Must have at least two basins for processing full plant flow. No standby tankage required	Must have at least two parallel basins.
Activated sludge – Aeration blowers	Must have sufficient capacity to meet oxygen demands with largest blower out of service	Must have sufficient capacity to meet oxygen demands with largest blower out of service
Activated sludge – Air diffusers	Air diffusion system must be designed so that largest section can be isolated without measurably impacting oxygen transfer capability	Plants with less than four independent aeration basins shall be designed with removable diffusers that can be serviced without draining the tank
Ultraviolet disinfection	Must be able to process 50% of plant flow with one (largest) unit out of service	Must have a minimum of two lamp banks per channel to facilitate cleaning or service while maintaining capacity
Aerobic and anaerobic digestion	At least two tanks required, but no standby tank required	Requires multiple units or alternative method of sludge processing/disposal, but no standby tank required
Sludge dewatering equipment	Centrifuges require backup unit, which can be uninstalled. No other redundancy listed for alternative dewatering systems	Not specified

N/A: Not specified

2.2.2 Statistical methodologies

Although there are many federal and state guidelines providing statistical methodologies for calculating the risks associated with wastewater effluent or potential pollution sources on the environment, there is little guidance for applying statistical methods to the wastewater process design itself.

In process design, various statistical methods are used to calculate design flows and loads. Design criteria from most of the standard texts dictate that 'maximum month' or 'peak' flows or loads be used. These are usually defined using a period of 5–10 years and are usually selected by calculating a percentile based on several years of data. Rigorous statistical methods are not typically applied to data screening or sample size collection.

Frequently, flows are analysed in more detail than loads, simply because there is often more data to work with. Multivariate regression analysis can be used to estimate collection system response to rainfall and soil conditions. This enables estimates of flow based on long-term rainfall event data as well as estimates of soil conditions based on time of year, frequency of serial rainfall events, and temperature.

2.2.3 Scenario analysis

An established method of accounting for variability and uncertainty is to run a steady state or dynamic model under several conditions. This one at a time type of scenario analysis tries to capture how the plant will operate under multiple conditions including conditions such as:

- High and low temperatures;
- High and low flows or loads;
- Seasonal permitting requirements; and
- Combinations of units in or out of service.

Additionally, in the preliminary design stage, multiple scenarios may need to be addressed that account for:

- Multiple population growth scenarios; and
- Multiple future permitting requirements.

The designer/engineer may choose to analyse the effect of other uncertainties that affect the design such as kinetic variables or wastewater characteristics that have not been well defined. The number of scenarios analysed is usually limited by the budget for planning and design and typically is focused on the most realistic or critical scenarios.

2.2.4 Mathematical modelling

The use of models to support the decision-making process has become common practice. The results obtained from modelling efforts must, however, be used judiciously, given the fact that plant design and operation remain vulnerable to imperfect data and to imperfect predictability of the system behaviour. When implementing models for design, the engineer must select plant-specific inputs to the model including detailed wastewater characteristics and biokinetic parameters. The operational envelope of the model-based design under evaluation can be tested by varying the values of the model parameters and influent characteristics.

With the development of sophisticated whole plant computer models, the standard design criteria can be challenged if the designer/engineer can convince regulators that the models reasonably predict plant performance. This will require detailed wastewater characterisation and/or pilot testing. In such a case, the engineer's and the regulator's judgement are generally used to determine the acceptable risk of applying modified criteria.

Model application for design typically requires that the designer/engineer identifies the most critical inputs and the most appropriate values for those inputs. The less critical model inputs can be fixed at a default value set by a knowledgeable model developer. The designer/engineer must select an appropriate methodology for determining flows, loads, and other model inputs that when combined do not result in an overly conservative design or a critically under-designed system. The designer/engineer must also be able to communicate the level of risk of critical design decisions to decision makers.

As computer processing power and speed has increased, interest in using Monte Carlo techniques has also increased. The Monte Carlo approach is attractive to treatment process design for several reasons including:

- The number of variables that can affect a design is high;
- The Monte Carlo method can account for covariance between variables;
- Sophisticated whole plant simulators are available that account for the interaction of multiple processes;
- Different unit processes may be affected in opposite ways by certain assumptions and the Monte Carlo method can test many assumption combinations;
- The Monte Carlo method enables the determination of peak, average, and minimum design requirements; and
- The Monte Carlo method enables the use of computing power to analyse multiple scenarios.

In academia, sophisticated statistical analyses are sometimes used for model calibration and process design. However, most of these methodologies have not been used outside of academia because they require significant compute resources, detailed data needs, as well as time and expertise to complete the analyses.

Chapter 3 discusses in detail the potential of incorporating mathematical models and statistical techniques for process design.

2.3 ADDRESSING SPECIFIC SOURCES OF UNCERTAINTY AND VARIABILITY IN CURRENT DESIGN PRACTICE

Even though not explicitly stated, design guidelines identify areas of uncertainty – in this STR called **sources of uncertainty** or **variability** (for definitions see Chapter 1, Box 1.2) – and assign safety factors to each one. The objective is to determine which of these sources of uncertainty are most important, and which have the biggest role in the decisions that need to be taken. As the project progresses and decisions are made, fewer sources of uncertainty need to be considered and the degrees of freedom in the design process are reduced.

2.3.1 Addressing sources of variability and uncertainty in flow and load determination

2.3.1.1 Use of historical information to develop design flows and loads

Good design practice is to use historical information as a component of the design basis for the facility. This includes actual facility data (e.g., raw influent flow and concentration data), population growth and projections, current and future zoning of the service area, past, present, and future capital improvement projects (e.g., infiltration and inflow improvements). For new facilities, data collected in nearby plants or in plants situated in similar catchments can be used.

Population growth projections are generally looked at in several different ways including, historical straight-line projections, traffic analysis zoning, and census projections. The evaluations are independently evaluated to determine future growth for flow projections. As with any analysis, the use of multiple data sets reduces the uncertainty in the evaluation. In most cases, the risk and uncertainty at this level is generally accepted by the owning entity.

Zoning changes and capital improvement projects in the service area can significantly impact the flow and characteristics of the wastewater conveyed to the treatment facility. Standard design practice is to consult the owning entity on future plans for the service area and to address changes expected during the planning life of the treatment facility during the preliminary design. This is typically done by adjusting historical facility data to account for these changes. In most cases, the risk and uncertainty regarding zoning changes is accepted by the owner.

Engineers will typically use peaking factors to account for the variability in flows and wastewater concentrations. These peaking factors can be developed by evaluation of historical data from flow meters or from empirical equations such as those provided in the 2014 edition of Ten States Standards (GLUMRB, 2014) that relate population to the hourly flow peaking factor.

Flow peaking factors are used to verify that facilities will perform at peak flow conditions as well as to confirm loading rates on unit processes such as clarifiers and tertiary filters. Mass loading peaking factors are commonly used to ensure performance and permit compliance.

It is the designer/engineer's responsibility to assign risks to the various peaking factors to size components such as bioreactor volumes, oxygen-delivery systems, clarifiers and filter surface area requirements, as well as chemical feed system requirements.

Risk and uncertainty in the use of facility data are associated with sample collection and analysis. Verifying the data quality procedures followed by the owning entity will reduce the uncertainty of the data sets. The designer/engineer further mitigates risk and uncertainty by evaluating the data and, for example, removing outliers. After the historic data set is modified, the design basis is modified further for the other factors described below. This is, of course, limited only to facilities already in service.

2.3.1.2 Use of per capita flows and loads

For existing facilities, the use of industry accepted per capita flows and loads, supplemented with population projections, can provide verification for facility design criteria. Significant discrepancies found in this verification step can warn of insufficient conservatism in the design. The designer/engineer needs to examine the cause of the discrepancy and should re-evaluate the design criteria if the discrepancies cannot be explained by a change in the service area or future capital improvements.

2.3.1.3 Screening of influent wastewater data

Analysis of the historical data is used to understand the influent wastewater characterisation. The designer will be more certain in his/her design if he/she is certain that the available influent data represent the true wastewater characteristics. Data evaluation techniques include data plotting, screening, flow- and mass balances, correlations, and the calculation of peak flow and mass loading factors. The specific methods used to evaluate the data vary widely as there is currently no commonly accepted best practice to do this evaluation.

2.3.1.4 Wastewater characteristics when data are not available

If wastewater characterisation data are not available, designers must make assumptions regarding the wastewater characteristics and loading patterns. These assumptions are often based on a combination of published information, information from surrounding facilities, and engineering judgement. The design then normally includes some safety factors because of the larger degree of uncertainty.

Wastewater characterisation also changes over time introducing uncertainty to the future plant performance. These changes can be attributed to the gain (or loss) of population, water consumption patterns, and industry. Wastewater characteristic changes should be estimated during design, especially when considering nutrient removal.

2.3.2 Addressing sources of uncertainty in unit process design

The following sections focus on how current practice addresses uncertainty in the design of unit processes. The continuous activated sludge system is used as an example.

2.3.2.1 Selection of design aerobic solids retention time

Perhaps the most common example of addressing uncertainty and variability in WRRF design practice today is the use of a safety factor when determining the aerobic solids retention time (SRT) for a nitrifying system. There are many variables related to both influent wastewater quality as well as operations that determine the system SRT needed to assure sufficient ammonia removal. These include parameters related to the growth rate of autotrophic organisms such as temperature and pH along with operational parameters such as the dissolved oxygen concentration and the ammonia concentration in the bioreactor. Other operational parameters, such as clarifier performance (solids leaving the plant) as well as waste activated sludge quantities, also impact SRT.

For example, in the ATV-DVWK-A 131 (2000) guidelines, the equation used for the calculation of the SRT includes a safety factor which takes into account: (a) potential variations of the maximum growth rate caused by certain substances in the wastewater, short-term variations and/or pH shifts, and (b) the variations of ammonium load. The guidelines suggest that the safety factor should be in the range of 1.4–1.8 (lower safety factors for higher population equivalents). Similar safety factors are included in most guidelines such as Metcalf and Eddy (Metcalf & Eddy Inc. *et al.*, 2014) and WRC (Ekama *et al.*, 1984) among others.

Due to the variability and uncertainty in both bacterial growth and plant operations, safety factors are used to ensure that washout of autotrophic organisms does not occur. The designer/engineer may also employ the use of a longer SRT to ensure that a target effluent ammonia concentration is met although increasing SRT does mitigate risks associated with nitrifier washout and high effluent ammonia concentrations, it does present additional challenges. Long SRT systems can be prone to filamentous bulking as well as high capital and operating costs, as a result of requiring more oxygen and larger tank volumes.

2.3.2.2 Selection of design sludge volume index

State-point analysis is commonly used to determine the horizontal surface area needed for secondary clarifiers and to determine underflow rates. Uncertainty in state point analysis outputs stems from uncertainty in the gravity flux curve and the Vesilind parameters. Additional limitations of the state-point analysis can be found in Henze *et al.* (2008).

Uncertainty relating to solids settling in secondary clarifiers typically results in a design sludge volume index (SVI) that is rather high. In order to mitigate the risk caused by the uncertainty of varying SVIs and operating conditions, the secondary clarifier is typically evaluated using multiple state-point analyses at varying design conditions to determine the performance of the clarifier under those different conditions (Henze *et al.*, 2008).

Empirical relationships between SVI and initial settling velocity (ISV) have been developed to be able to generate solids-flux curves based on SVI and mixed liquor suspended solids concentrations (Daigger, 1995).

This helps the designer/engineer to quickly evaluate clarifier conditions without knowing facility specifics. To mitigate the risk associated with these empirical relationships, column testing can be performed using the facility's mixed liquor to develop the solids-flux curve for that individual system, but allowances will still be needed for varying conditions.

The WRC guidelines (Ekama *et al.*, 1984) include an explicit safety factor that is used to multiply the estimated area of the secondary settling tank. The area of the secondary clarifier is estimated as a function of peak wet weather flow, mixed liquor suspended solids concentration, the recycle ratio, and $SSVI_{3.5}$ using an empirical equation that has been derived based on flux settling parameters measured at 30 plants in the UK. The calculated area is then multiplied by a safety factor of 1.25.

2.3.2.3 Selection of design denitrification rates

Uncertainty relating to the denitrification rate in nitrogen removal facilities is a function of several items: temperature, pH, dissolved oxygen carry-over from aerated zones, use of light aeration in anoxic zones to maintain solids in suspension, availability of readily biodegradable (fermentable) COD (S_B) in the anoxic zone feed and hydrolysis of particulate biodegradable COD (X_B) to S_B. The designer/engineer can account for variations in temperature by determining minimum temperature requirements and sizing the reactor accordingly. Alkalinity balances can be performed to determine if pH is impacted and, if needed, alkalinity feed systems can be added. The impact of dissolved oxygen can be accounted for during design by providing tapered aeration, real-time aeration control and/or de-oxygenation zones for mixed liquor internal recycles.

Uncertainty relating to the denitrification rate in nitrogen removal facilities is typically handled by the appropriate sizing of the anoxic zone. For example, in the ATV-DVWK-A 131E (2000) design guidelines, the size of the anoxic tanks has to satisfy the recommended ratio of the anoxic to total volume of the bioreactor. Ratios of less than 0.2 or greater than 0.5 are not recommended.

In the WRC guidelines, the anoxic tank volume is derived from the aerated section volume. The volume of the aerated sections of the bioreactor is calculated as a function of SRT and the maximum specific growth rate of the nitrifying organisms. The recommended values for the un-aerated to the total bioreactor volume are presented graphically and indicate that the ratio should not be larger than 60%.

Historically, the equations used for the sizing of the anoxic zones have been proven to be conservative, alleviating the risk involved in meeting effluent total nitrogen concentrations. In the event that the design engineers feel that the risk has not been adequately addressed, they often choose to add tertiary treatment.

Variations in readily biodegradable COD (S_B) in the influent wastewater cannot be mitigated by the designer/engineer or operations staff. The uncertainty in this parameter is exacerbated by the fact that most treatment facilities do not perform S_B measurements, which require either respirometry or a combination of physical chemical analyses (Choubert *et al.*, 2013; Melcer *et al.*, 2003). This component of the influent waste stream is vital for both biological phosphorus removal and denitrification.

The uncertainty related to S_B for denitrifying systems is generally accounted for in two ways. The designer/engineer might rely on either empirical equations (or use an uncalibrated process model) to calculate a denitrification rate, or might use a value of the hydraulic retention time based on a rule of thumb to directly calculate anoxic volume.

Several empirical equations and curves have been developed to determine denitrification rates for pre- and post-anoxic zones. The most prevalent equation for pre-anoxic zone sizing was published by Burdick *et al.* (1982), which relates the F/M ratio to the denitrification rate.

In addition to empirical relationships, the size of the anoxic zones is frequently determined with a simulator.

2.3.2.4 Selection of dissolved oxygen concentration in bioreactors

Designers typically will select design dissolved oxygen concentrations for varying design conditions (average day, maximum day, etc.) to ensure that there is adequate oxygen available for oxidation of carbonaceous and nitrogenous matter. Historically, activated sludge plants have been designed to operate at a dissolved oxygen (DO) concentration of 2 mg/L as a means to account for uncertainty in aeration demand due to variability in wastewater strength and temperature.

2.3.2.5 Selection of design oxygen transfer efficiency

The designer/engineer often assumes several key parameters that have large impacts on the sizing of air delivery systems in wastewater treatment. These include the alpha value, diffuser fouling factor, the standard oxygen transfer efficiency (SOTE) for diffused air systems and the standard aeration efficiency (SAE) for mechanical surface aeration systems.

Alpha values are typically prescribed in industry-accepted literature based on the method of aeration being employed. Field testing can also be done to determine this number. The standard oxygen transfer efficiency is the percentage of the oxygen transferred into the mixed liquor from the overall amount of oxygen delivered at standard conditions. This number varies based on the diffused air method used, as well as the depth of the diffusers. The standard aeration efficiency (measured in kg/kw-hr or lb/hp-hr) is generally provided by the surface aerator manufacturer, and is often found to be unrealistic in actual applications. These values have been scrutinised over the years and found to be overly aggressive. Field testing done by third parties has indicated SAE values lower than the typical claims of the manufacturer.

The oversizing of air systems can be problematic from both a capital investment standpoint as well as from an operational standpoint. Providing too much air will impact the biology of the mixed liquor potentially causing poor settling. For facilities employing nutrient removal, high dissolved oxygen concentrations in recycle flows can impact the performance of fermentation and anoxic zones.

These risks are typically addressed by sizing air systems with adequate turndown through the use of multiple units and/or use of variable frequency drives to ensure that sufficient air is provided at both the minimum and maximum design condition. Automatic control systems to control the speed on blowers can also be employed to ensure that the proper amount of oxygen is provided to the system. Risk can further be mitigated through field oxygen transfer testing to determine actual field transfer conditions.

2.3.3 Addressing uncertainty via effluent permit selection

The following sections discuss how uncertainty in WRRF performance can be taken into account by selecting more conservative effluent permits both as a permit writer as well as a design engineer. To illustrate the point, an example for the USA legal framework has been included.

2.3.3.1 Effluent limits

In the United States, the Environmental Protection Agency (USEPA) Clean Water Act (USEPA, 1987b) requires that any point source discharge to a navigable water body be permitted through the USEPA or a State with delegated authority under the National Pollution Discharge Elimination System (NPDES) programme. The Clean Water Act specifies limitations using 'best available current technology' to issue technology-based effluent limits.

Permitting authorities are required to add more stringent water-quality-based standards for impaired waters. The total maximum daily load (TMDL) programme was instituted as part of the Clean Water Act to identify and determine point and non-point source reductions to impaired water bodies with the intent that the water met the applicable designated uses.

The Florida Administrative Code (Fla. Admin. Code, 2013) explicitly lists effluent limits for discharge to ocean outfall, deep well injection, reclaimed water, and for surface and groundwater discharges that do not have water-quality-based standards. This removes all issues of dealing with risk and uncertainty from the wastewater permit writer.

The requirement for water-quality-based standards indicates that the surface water is impaired and has an approved TMDL. The TMDL is developed through water quality modelling done by those other than the wastewater permit writer. The data used in the water quality modelling is either real or generated by the water quality modeller (water quality modellers use only a margin of safety factor to account for uncertainty). When multiple discharges occur within a discharge segment the permit writer must consider the waste load allocation (WLA). Non-points and natural sources are included as a load allocation (LA). The modeller uses the following formula to develop a TDML for a specific parameter; TDML = WLA + LA + MOS. Stakeholders (those contributing to the impaired water body) can provide public input and data to assist in the development of the TMDL.

Once the TMDL is established and approved by the USEPA, the wastewater permit writer incorporates it into the permit. The wastewater or stormwater permit writer does not take on any risk in issuing this numerical limit as it has been established and approved by others. When multiple discharges occur within a discharge segment the permit writer must consider the WLA as it was utilized during the modelling process.

2.3.3.2 Selection of effluent design criteria

Facility design is based upon meeting a numerical effluent limit in order to meet a permit requirement.

The designer/engineer normally employs a lower target effluent concentration in the process design, as compared to the permit limit, to account for uncertainty. For high rate (non-nitrifying) facilities requiring only BOD and TSS removal, assuming lower BOD and TSS values in the effluent do not significantly impact facility sizing if guidelines, such as selection of SRT to washout nitrifiers and clarifier loading rates, are followed.

Larger impacts are common where nutrient removal is required and the designer/engineer accounts for uncertainty by utilizing a design effluent ammonia, nitrate, or total phosphorus value lower than the effluent limit. Modelling with Monod kinetics has shown that lower substrate concentrations decrease the growth rate of the organism. For nitrifying bacteria, use of a lower than required substrate concentration will result in a larger bioreactor. This relationship is not linear and, therefore, a slightly modified effluent target concentration can significantly impact a modelled growth rate and bioreactor size.

Risk mitigation options depend on the effluent compliance period of the facility. For example, if the plant has a very low phosphorus limit over a short averaging period (e.g., 7-day average or monthly average), significant risk mitigation methods may be warranted to address the issue of even a 'small' excursion causing a permit violation.

Risks over meeting total phosphorus limits are sometimes mitigated by the designer/engineer using a lower effluent total phosphorus value, which often requires the use of increased metal salts during operation. This may ensure compliance with effluent phosphorus limits at the cost of additional operational costs for the metal salt, a significant increase in solids production, and potentially detrimental effects on the pH in the process.

2.3.4 Summary of uncertainty analysis methods in current practice

Table 2.2 summarises the methods used in practice during design to address key sources of uncertainty and variability. Most of the engineering decisions are made during this phase and thus it is important to be able to quantify the associated uncertainty.

Table 2.2 Summary of methods used in practice to address key sources of uncertainty and variability.

Source of Uncertainty	Uncertainty or Variability	Risk	How Practice Addresses Risk
Influent flows and mass loads	Rate of increase of flow and concentration Peak flow and loading events Correlation between flow and load Variability of historical flows and loads Data accuracy	Underestimating flows and loads Overestimating flows and loads Ignoring correlations or lack of correlations between flows, loads temperature, and discharge requirements. Changes in flows and loads due to changes in population or service area make-up	Use of historical information to determine population growth rates and peaking factors Verification of historical data by using per capita mass loads and flows Screening of data and omission of outliers Use of flow and load peaking factors for design of hydraulic elements and unit processes
Characterisation of the wastewater components	Consistency of fractionation over project planning period Lack of long-term historical data to measure COD fractionation	Overestimation of nutrient removal and/or sludge quantity Underestimation of nutrient removal and/or sludge quantity	Sensitivity analysis Practical checks of process models vs. traditional design criteria
Aerobic SRT	pH and DO control Variability and correlation of pH, temperature Nitrification rate Plant operations	Washout of autotrophic organisms Effluent ammonia concentrations exceeded Bulking sludge	Washout SRT safety factor
Design SVI	Plant upsets Plant operations	Poor settling mixed liquor Clarifier failure	Use of selectors Clarifier safety factors Percentile evaluation on historical SVI

(*Continued*)

Table 2.2 Summary of methods used in practice to address key sources of uncertainty and variability (*Continued*).

Source of Uncertainty	Uncertainty or Variability	Risk	How Practice Addresses Risk
Nutrient Uptake rate	Variability of pH, DO, temperature in wastewater Abundance of readily biodegradable organic matter in wastewater	Exceeding permit requirements Oversizing of anoxic zones which can result in phosphorus release in bio-P systems	Sizing reactor for permit condition requirements Use of empirical equations for anoxic zone sizing for facilities not employing bio-P Addition of supplemental carbon source Addition of post-denitrification capabilities Process simulation
Process air system design	Design condition (max day/max week) Oxygen inputs upstream of system (cascades)	Inadequate air at high demand conditions Overdesign at low flow/start-up conditions Affects nitrifier growth rate Affects system microbiology (filamentous organisms)	Provide equipment available to meet peak demand Have flexibility to turndown oxygen delivered Dissolved oxygen control via instrumentation
Effluent design criteria related to the effluent permitting requirements	Accuracy of water quality models predicting receiving water quality. Variations in receiving water quality and flow Seasonal permit limits Accuracy in models in predicting effluent quality Future changes in regulations Facility operations	Permit compliance	Use of lower design effluent limits as compared to permit requirements

2.4 IMPLICATIONS OF CURRENT PRACTICE ON DEGREES OF FREEDOM IN ENGINEERING DECISIONS

Depending on jurisdiction, design approaches can vary from highly prescriptive to very open, resulting in varying degrees of freedom in the decision-making process.

The use of strict industry standards in design (similar to strictly following a recipe in a cookbook) automatically reduces the degrees of freedom in the decision-making process. If one assumes that decisions on loads and effluent requirements have been taken and the industry standard is to be strictly followed, then the design becomes an automatic procedure that does not require any decision making. This approach 'buries' uncertainty, which is not seen by the stakeholders, and normally increases project costs significantly.

In most cases though, even when industry standards are purported, engineering judgment is still required and parameters that differ from the default values might be used by the designer (industry standard is used as a guideline). In this case, the designer/engineer will need to select values for the design inputs such as safety factors for nitrification or a sludge volume index to mitigate his/her risk.

When no industry standard is purported, and the engineer is free to choose the design methodology, the degrees of freedom increase dramatically and by extension so do the sources of uncertainty to be considered. In this case, the engineer is able to make decisions on the selection of a guideline or a process model as well as the values for all of the design inputs.

It is evident that in the first case (strict adherence to a guideline), no competition in the design can arise and both the design engineer and the owner are legally protected in case of failure by having followed a pre-selected state-of-the-art procedure. However, in this case there is little possibility to seek out opportunities and to look for optimised or competitive solutions. Also, the choice of technologies and configurations will be restricted which may lead to non-optimal solutions. At the other extreme, in case 3, the encountered uncertainties may give rise to both a risk of failure as well as opportunities that arise from competition. It could be argued that the industry is moving from case 2 to case 3 where the consortium needs to cover for risk of failure but is also able to reap the benefits from innovative ideas. In case 3, the need for structured appraisal of sources of uncertainty and variability gains importance.

2.5 SUMMARY

Risk discussed in this chapter is associated with uncertainty in the design process. Uncertainty during the design process results in (usually) the selection of conservative assumptions for the basis of design. This uncertainty is addressed through the use of statistical methods that discard outliers in data, the use of safety and peaking factors in design, the use of effluent criteria that are lower than permit standards, and generally accepted methods for determining nutrient uptake rates and oxygen requirements. Each of these decisions impact both the operational flexibility of the facility and the construction and operational costs.

Ultimately, designers/engineers, owners, contractors, and regulators, need to understand the interactions between making conservative assumptions in design and the impacts of those assumptions on the project lifecycle cost. The cost of providing conservative water quality standards, coupled with the safety factors used during the design process, will most likely not cause a linear increase in project cost but rather an exponential increase depending on the conservatism used for major design decisions. Future work that could determine the overall 'conservatism' contained in a facility arising from all decisions taken from the creation of water quality standards all the way through to process design would be very valuable.

REFERENCES

ATV-DVWK-A 131E. (2000). Dimensioning of Single-Stage Activated Sludge Plants, German ATV-DVWK Rules and Standards. German Association for Water, Wastewater and Waste, Hennef, Germany.

Burdick C., Refling D. and Stensel H. (1982). Advanced biological treatment to achieve nutrient removal. *Journal Water Pollution Control Federation, USA*, **54**(7), 1078–1086.

Choubert J. M., Copp J., Gillot S., Melcer H., Morgenroth E., Rieger L., Rönner-Holm S., Shaw A., Spérandio M. and Sørensen K. (2013). Rethinking wastewater characterisation methods for activated sludge systems – a position paper. *Water Science and Technology*, **67**(11), 2363–2373.

Daigger G. T. (1995). Development of refined clarifier operating diagrams using an updated settling characteristics database. *Water Environment Research*, **67**, 95–100.

Doby T. A. (2004). Optimisation of Wastewater Treatment Design under Uncertainty and Variability. PhD thesis, North Carolina State University, North Carolina, USA.

Ekama G. A., Siebrits I. P., Marais G. V. R., Pitman A. R. and Smollen M. (1984). Theory, Design and Operation of Nutrient Removal Activated Sludge Processes. University of Cape Town, City Council of Johannesburg and CSIR, WRC Report No TT16/84.

Florida Administrative Code. (2013). Discharge to Surface Waters – Ocean Outfalls. Department of Environmental Protection. ID: 13679927.

Grady C. P, Daigger G. T., Love N. G. and Filipe C. D. M. (2011). Biological Wastewater Treatment, 3rd edn. IWA Publishing, CRC Press, London, UK. ISBN13: 9780849396793.

Great Lakes Upper Mississippi River Board (GLUMRB). (2014). Recommended Standards for Wastewater Treatment Facilities (Ten State Standards). Health Research, Inc. Albany, NY, USA.

Henze M., van Loosdrecht M. C. M., Ekama G. and Brdjanovic D. (2008). Biological Wastewater Treatment: Principles, Modelling and Design. IWA Publishing, London, UK. ISBN13: 9781843391883.

Lawrence A. W. (1971). Application of process kinetics to design of anaerobic processes. In: Anaerobic Biological Treatment Processes, Chapter 9, *Advances in Chemistry*, **105**, 163–190, DOI: 10.1021/ba-1971-0105.ch009.

Lawrence A. M. and McCarty P. L. (1970). Unified basis for biological treatment design and operation. *Journal of the Sanitary Engineering Division*, **96**, 757–777.

Melcer H., Dold P., Jones R., Bye C., Takacs I., Stensel D., Wilson W., Sun P. and Bury S. (2003). Methods for Wastewater Characterization in Activated Sludge Modeling. WERF Project 99-WWF-3. Published by Water Environment Federation, Alexandria VA and IWA Publishing, London, UK.

Metcalf & Eddy Inc., Tchobanoglous G., Burton F.L., Tsuchihashi R. and Stensel H.D. (2013). Wastewater Engineering: Treatment and Resource Recovery, 5th edn. New York, NY: McGraw-Hill Professional.

Palmer T. M., Ross M. C. and Jank B. (2003). Efficient Redundancy Design Practices. International Water Association (IWA), London, UK.

Russell D. (2019). Practical Wastewater Treatment, 2nd edn. Wiley and Sons Publishing, Hoboken, NJ, USA.

United States Environmental Protection Agency (USEPA) (1974). Design criteria for mechanical, electric and fluid system and component reliability EPA-430-99-74-001, Office of Water Program Operations, Washington, DC.

United States Environmental Protection Agency (USEPA) Clean Water Act. (1987a). Federal Water Pollution Control Act. https://www.epa.gov/sites/production/files/2017-08/documents/federal-water-pollution-control-act-508full.pdf

United States Environmental Protection Agency (USEPA). (1987b). Phosphorus Removal Design Manual, EPA 625-1-87-001, Cincinnati, OH, USA.

United States Environmental Protection Agency (USEPA). (1993). Nitrogen Control Manual, EPA 625-R-93-010, EPA, Office of Water, Washington, DC, USA.

United States Environmental Protection Agency (USEPA). (2000). Nutrient criteria technical guidance manual: Rivers and streams, 822-B-00-002, EPA, Office of Water, Washington, DC, USA.

Virginia DEQ. (2008). Virginia's Sewage Collection and Treatment Regulations (form 9VAC25-790).

Wahlberg E. (2004). WERF/CRTC Protocols for Evaluating Secondary Clarifier Performance. Water Environment Research Foundation, Alexandria, VA, USA.

WEF MOP-29. (2005). Biological Nutrient Removal (BNR) Operation in Wastewater Treatment Plants. Water Environment Federation Manual of Practice (MOP) No29. Water Environment Federation, Alexandria, VA, USA.

WEF MOP-8. (2017). Design of Water Resource Recovery Facilities, 6th edn. Water Environment Federation Manual of Practice (MOP) No8, McGraw-Hill, New York, NY, USA.

Chapter 3

Incorporating uncertainty analysis into model-based decision making – opportunities and challenges

3.1 INTRODUCTION

There is increasing consensus among water reclamation professionals that the predictive power of models is a critical component of plant design and operation. However, models and simulators used by designers and operators are yet to systematically incorporate methods for the evaluation of the uncertainty associated with design and operations. Such approaches have the opportunity to assess the upset resilience of individual processed and the system as a whole.

This chapter presents a general discussion on the implications of uncertainty when using process models for design and optimisation of treatment plants. The discussion highlights both the potential opportunities for explicit accounting of uncertainty as well as open questions that need to be dealt with before such approaches can be integrated into daily design practice.

3.2 INCORPORATION OF SAFETY IN CURRENT MODEL-ASSISTED DESIGN

Common practice for the design of a treatment plant is to use design guidelines. In some cases, the designer/engineer will then run a steady state or dynamic simulation with a process-based biokinetic model and if the predicted effluent concentration is (well) below the effluent requirements, then the design will be judged as appropriate.

In some cases, when a model is directly used to obtain a design, uncertainty and variability are accounted for in several ways as discussed in Chapter 2:

- Use of a higher influent load than the design load;
- Design to more stringent effluent requirements (e.g., to 70% of the effluent limit concentration);
- Choice of conservative values for process parameters (e.g., reduced maximum growth rates, increased sludge volume index);

Editor(s): Evangelia Belia, Lorenzo Benedetti, Bruce Johnson, Sudhir Murthy, Marc Neumann, Peter Vanrolleghem and Stefan Weijers
doi: 10.2166/9781780401034_0033

- Increase of obtained design variables (e.g., multiply a resulting design volume with a safety factor of 1.5);
- A combination of the above.

An alternative would be to use a model-assisted approach where the designer/engineer tries to directly quantify the uncertainty associated with the different model factors (parameters and inputs). Uncertainty propagation can then be used to quantify how the plant performance predicted by the model is affected. Design variables (such as tank volume) can then be iteratively modified until the probability of compliance reaches a specific value, for example, 95% or 99%. This probability ties into the resilience of the process and the overall treatment system.

3.3 OPPORTUNITIES OF EXPLICITLY CONSIDERING UNCERTAINTY AND VARIABILITY

One can view the use of safety factors in a design guideline as an implicit way of dealing with uncertainty. By adding a margin of safety, various sources of uncertainty are accounted for simultaneously. The advantage of this approach lies in its simplicity. Historically developed safety factor approaches have withstood the test of time and are often widely accepted as industry standards. However, guidelines are not available for all configurations or new technologies and thus may limit innovation. This is especially true when highly integrated systems are being conceived (water resource recovery facilities (WRRFs) with many feedback loops due to sophisticated control or interconnected systems, e.g., sewer-WRRF-river). Recent developments in simulation-assisted approaches means that the behaviour of such complex systems can be described.

The challenge when using a process-based model is how to appropriately account for uncertainty, that is, how to appropriately translate the safety factor approach of a guideline. The designer/engineer may explicitly acknowledge the uncertainties present in the modelling exercise in different ways. One possibility is to document how each uncertain input value was determined, to provide the rationale for the decisions made for each of the uncertain inputs. Another possibility would be to express the uncertainty by using quantitative probability distributions for model inputs and model parameters.

The main practical advantage of explicit approaches is increased transparency regarding uncertainties (Flyvbjerg *et al.*, 2003). It is expected that such methods can be refined by conducting post-project audits during which the assumptions made in the design phase are compared with the performance of the built plant and the causes of any discrepancies are identified.

3.4 SCOPE AND LIMITATIONS OF MODELS

Models by definition are not exact replicas of real-world systems. The necessary simplifications inevitably lead to the introduction of uncertainty. When using a model, the model scope needs to be considered. For example:

- Which phenomena were included, and which ones were not included?
- What is the range of values for model inputs and parameters for which the model is expected to give adequate results?

3.4.1 Evolution of wastewater treatment modelling

The scientific development of activated sludge models (ASMs) reflects the improvements in the understanding of the fundamental microbiological transformation processes occurring in biological wastewater treatment and the current ASMs are widely applied in engineering practice as state-of-the-art models and used to predict plant performance.

3.4.2 Desirability criteria for models

The following features are often used to judge the desirability of a model (e.g., Reichert, 2009):

- **Causality**: The model represents the relevant cause–effect relationship for the system response at the required level of resolution.
- **Universality**: The model structure and as far as possible the model parameter values should be transferable to a similar system.
- **Predictive capability**: The model should remain valid for the extrapolation of external influence factors to ranges required for predictive use.
- **Identifiability**: Unknown parameter values should be identifiable with the available data. This means that a model can be fitted (i.e., parameters can be identified) when applying a fitting (optimisation) algorithm. When this is not the case, prior knowledge of non-identifiable parameters must be available.
- **Simplicity**: The model should be as simple as possible.

With respect to the five desirability criteria listed above, an attempt to classify the importance of these features to ASMs is: causality (high), universality (medium–high), predictive capability (medium–high), identifiability (see Box 3.1) (low), simplicity (medium). The strength of the ASM suite lies in characterising the microbial processes. However, when trying to emulate full-scale systems, other factors such as hydrodynamics, mass-transfer (such as oxygen transfer during aeration), varying sludge settling characteristics, precipitation chemistry, sensors and actuators, equipment reliability may become equally important or even more important. The following example shows the relationship between WRRF model limitations and sources of uncertainty.

3.4.3 Example of wastewater treatment plant model limitations

Consider the following scenario: An engineer is given all the details of the configuration of a plant and is then asked to model the (expected) effluent time series of the plant using one year of measured influent data (Figure 3.1). For this example, it is assumed that the measurements do not contain errors and are representative of the true values.

Assuming that the model is representative of the behaviour of the full-scale plant, it is reasonable to expect the modelled and the measured effluent time series to be similar. However, if the plant effluent data include an effluent limit exceedance due to a toxic spill from an industrial source, resulting in inhibition of microorganisms, this will not typically be reflected in the modelled time series, as such processes are not captured in the ASM-based models.

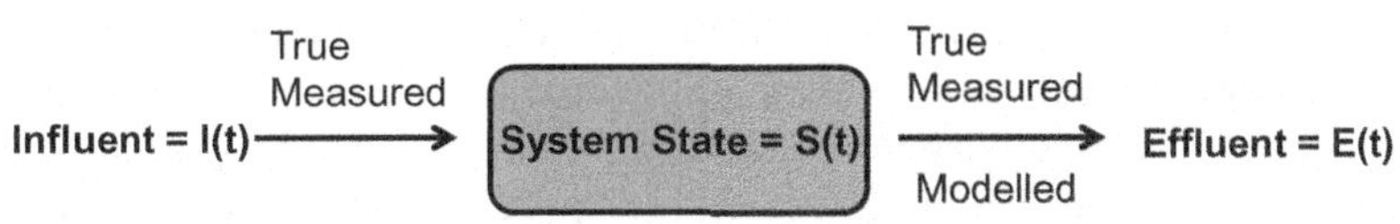

Figure 3.1 Influent, effluent and system state as functions of time. The effluent can be modelled using the measured influent and a model representation of the system state.

The comparison of the predicted (modelled) and the measured effluent time series demonstrates what the model does and does not account for. It identifies the processes that are not included in the model domain.

This simple example demonstrates the challenge faced when using standard ASMs for the prediction of WRRF behaviour. Because models do not **perfectly** emulate real-world plant behaviour, there exist processes and events, such as equipment failures, that impact plant performance but are not captured in the models.

It is therefore essential for the designer/engineer to determine which processes are crucial for each model-based design project. If additional processes or elements (e.g., equipment models) need to be included, the models need to be expanded (e.g., Rosen *et al.*, 2008). However, even in cases where model expansion is required but not possible, models can be used very effectively to compare alternative configurations with the same assumptions.

3.5 WHAT DON'T WE KNOW ABOUT DEALING WITH UNCERTAINTY?

Although there are many ways to include explicit descriptions of uncertainty in model-assisted designs, there remain several challenges. A selection of challenges that the authors believe to be relevant are listed in the following sections.

3.5.1 How conservative are we with the safety factor approach?

To evaluate how conservative the safety factor approach is, we need to answer the question: what do safety factors account for? This question is difficult to answer. On the one hand, safety factors have been developed over time, and the rationale and the data that were used to determine their values are not always known. On the other, it is very difficult to assess the actual quality of a design: ideally one needs the designed plant running at design conditions to determine if it is over- or under-designed. However, by the time that the design conditions are reached the plant has typically undergone substantial changes compared to the start-up configuration.

Deviations from the original planning assumptions could quite easily be detected in post-project audits: for example, number of person equivalents connected to a WRRF. Deviations from design assumptions, for example, critical SVI or growth rates could be detected by long-term monitoring of such parameters. Therefore, long-term post-project audits might be a valuable tool to improve design approaches. Long-term monitoring of influent and effluent quality could provide information for developing more robust design procedures (e.g., Bott & Parker, 2011).

Also, inter-guideline comparisons and the comparison of guidelines with ASM-based process models (Corominas *et al.*, 2010) can help quantify margins of safety inherent in design approaches and enable the exploration of aggregated safety factors.

3.5.2 How to move from guidelines with the safety factor approach to probabilistic model-assisted design?

Often biokinetic models are used to verify designs obtained from guidelines or standards. It is not common practice to design a facility solely with the use of a biokinetic model especially for greenfield plants. There are several reasons for this. In some cases, the engineer is legally protected if he/she uses a 'recognised' design guideline or standard. Model-assisted design may transfer liability to the engineer. Even if there are no liability issues with respect to model-based design, an engineer will need to decide where to add safety and how much to add. In many cases, biokinetic models are much easier to employ for plant upgrades where prior knowledge exists on influent and effluent characteristics, process operations, or verifiable model parameters. In these cases, a probabilistic model-assisted design can more easily be used as a primary approach for design, than in the case of a green field project. Under such conditions innovative technologies that are not already included in guidelines or standards can be incorporated.

3.5.3 Determination of prior uncertainty ranges

Two major issues to address within a probabilistic framework include which elements to select as uncertain and, secondly, how to quantify the uncertainty surrounding them. It is common practice for prior uncertainty ranges to be obtained from experts who have experience in determining typical values. Many models and simulation software also provide helpful prior uncertainty ranges.

3.5.4 Parameter (uncertainty) estimation in systems with poor identifiability

Available data, together with models can be used to estimate parameters and their uncertainty by using parameter estimation techniques. However, poor identifiability (see Box 3.1) of ASM-based models remains an issue (see also Box 3.2).

BOX 3.1 PARAMETER IDENTIFIABILITY

Parameters are identifiable when model fitting (optimisation) algorithms are able to find best estimates for the model parameters. For ASM-type models this is usually not the case due to the high number of parameters which one aims to identify and due to the lack of sufficient data.

BOX 3.2 FREQUENTIST VS. BAYESIAN PARAMETER ESTIMATION

In **Frequentist** parameter estimation, the models are confronted with observed data to obtain parameter estimates. Hereby, random observation errors in the data are mapped to uncertainty about the parameters. As wastewater treatment plant models are typically over-parameterised, optimisation (fitting) algorithms cannot find a unique solution due to compensation (technically termed non-identifiability, see Box 3.1). For example, increasing the value of one parameter can be compensated by decreasing the value of another one. Therefore, most parameters are set to default values (taken from literature) and only a few are estimated with statistical techniques. This leads to biased estimates (i.e., dependent on where the other parameters were fixed).

An alternative is to use a **Bayesian** framework where information from literature and expert knowledge can be combined to define prior value ranges for all the parameters. Then, the probabilistic model is confronted with the data and the parameter ranges are updated, typically narrowed. This framework does not require identifiability of the parameters. However, this framework requires the elicitation of 'inter-subjective' (i.e., experts must agree) prior parameter ranges. Also, in this framework, the parameter updating procedure is computationally expensive which may still be a limiting factor for dynamic WRRF models.

Often, the data does not provide enough information to enable a statistical estimation of uncertain parameters. As a result, engineering practice often fixes most of the model parameters at default values and just a few are estimated/calibrated. As a consequence, it is important that the estimated parameter values as well as their uncertainty estimates be assessed in an informed manner. One approach for obtaining uncertainty ranges is by extracting probability distributions for an influent characteristic or model parameter using performance data at brownfield locations (Alikhani *et al.*, 2017; Sharifi *et al.*, 2014).

3.5.5 How to adequately deal with biokinetic model structure uncertainty?

Uncertainty about the biokinetic model structure remains a difficult issue. Conducting statistical inference in the presence of model structure error, leads to biased estimates of the parameter values and unreliable uncertainty assessment (Neumann & Gujer, 2008; Villez *et al.*, 2020). This is especially critical for practising engineers as they cannot be expected to modify predefined bio-kinetic model structures (e.g., ASM). Nevertheless, several model structures are now available in most commercial simulators. In engineering projects, model structure selection often depends on the key process being simulated or the effluent parameter(s) associated with a permit limit.

3.5.6 Full-fledged probabilistic model-based design

When considering the replacement of a design guideline with a probabilistic WRRF simulator it is important to identify which real-world phenomena are included. If the model is expected to be a true emulator of the future WRRF, then current models would need to be significantly enhanced with models that account for operational behaviour of sensors, actuators and other equipment. They would need to be able to re-produce toxic spill events, inhibition events, bulking and foaming events, and operator failures, among other things. If these aspects are not covered by the simulator, alternative ways need to be found to take them into account. Such aspects should be clarified by modellers and process engineers by adding a disclaimer on which kind of processes are included in the model and which ones are not.

3.6 HOW CAN WE CURRENTLY ACCOUNT FOR VARIABILITY AND UNCERTAINTY?

3.6.1 Accounting for variability

Whereas accounting for temporal variability is the central aspect of dynamic modelling, spatial variability has until recently only been coarsely resolved using compartmental models such as tanks-in-series.

3.6.1.1 Temporal variability

Accounting for temporal variability include the use of probability distributions, dynamic modelling (multivariate) time-series analysis, and influent generators.

Probability distributions can be used to characterise the variability of dynamic variables such as flows, concentration or loads. This approach is also useful when using a steady-state solution of the model: for example, when describing average monthly behaviour, the influent concentrations and flows can be sampled from the probability distributions to capture meaningful scenarios (Bixio *et al.*, 2002; Mc Cormick *et al.*, 2007).

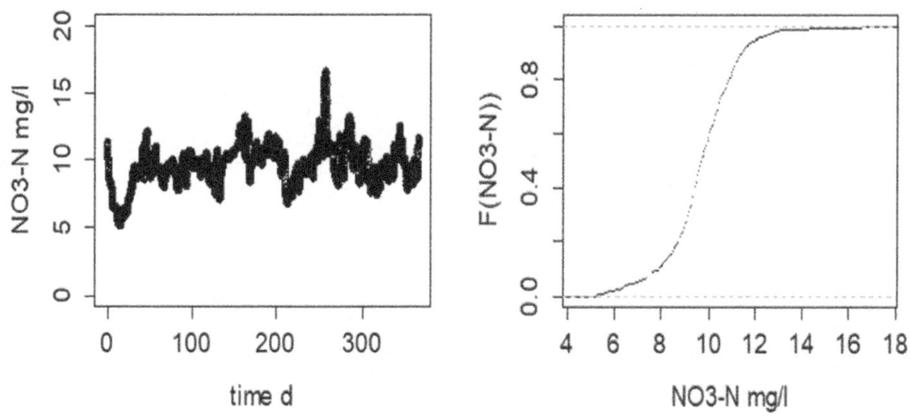

Figure 3.2 365 days of NO_3-N effluent concentration data (sampling interval: 1.2 hours). Left: Time series. Right: The corresponding empirical cumulative density distribution.

In addition, empirical cumulative distributions are often used to characterise plant performance. They condense the information contained in a time series and can, for example, extract the frequency of exceedance of effluent concentration limits. An advantage of directly using time-series analysis over distributions, is that temporal dependence (auto-correlation) is appropriately and explicitly accounted for. A typical time-series analysis identifies:

- trends;
- periodic phenomena;
- autocorrelation.

An example is given in Figure 3.2 for one year of nitrate effluent data with the original time series in the left panel and the corresponding cumulative distribution in the right panel. The y-axis in the cumulative distribution quantifies the percentage of time that the concentration is below the value on the x-axis.

To account for correlation between variables (cross-correlation) the same procedure can be followed with multivariate techniques. Dynamic simulators capture how dynamic influents affect the state variables of the system and predict a dynamic effluent profile from which desired statistics can be extracted.

If synthetic time series are required that represent future load scenarios, then influent generators can be used (Gernaey *et al.*, 2011; Martin & Vanrolleghem, 2014). Influent generators are typically either based on (stochastic) catchment models or are derived from black-box models that are calibrated with historic time series.

3.6.1.2 Spatial variability

Concerning the description of spatial variability, the rapidly growing computational fluid dynamics (CFD) field enables the investigation of spatial phenomena at high resolution (e.g., Alvarado *et al.*, 2013; Gresch *et al.*, 2011; Rehman *et al.*, 2017). Such analyses are critical for multiphase systems (e.g., settling), systems with spatial heterogeneity (e.g., anaerobic digestion) or systems that need to guarantee a certain contact time (e.g., disinfection). To decrease the computational burden, methods have been developed that enable the translation of a CFD model to a compartment model (Gresch *et al.*, 2009).

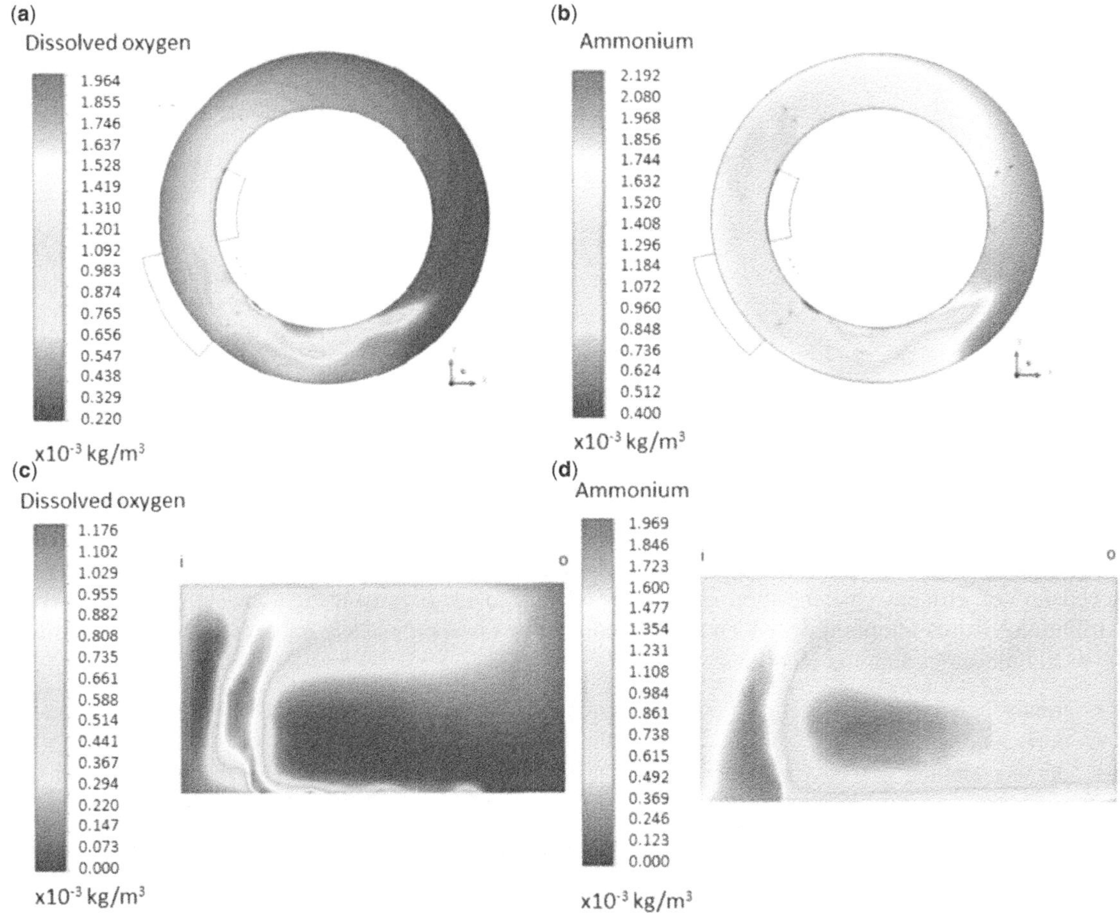

Figure 3.3 2-D concentration profiles in a reactor (Rehman *et al.*, 2017).

An example of a typical CFD model output illustrating the distribution of mean residence time in a reactor is shown in Figure 3.3.

3.6.2 Accounting for uncertainty

3.6.2.1 Uncertainty related to design scenarios

When planning a treatment plant, significant uncertainty is associated with defining the appropriate design loads. This uncertainty can be accounted for by applying foresighting tools such as scenario analysis techniques. These techniques enable multiple possibilities of future loads or other requirements to be accounted for (e.g., Dominguez *et al.*, 2009). They often involve expert interviews and participatory methods. A systematic use of such techniques within the water resources field is not yet widespread, but is increasingly an imperative, especially in cities where urbanisation is rapid or, in service areas where changes to wastewater characteristics are anticipated due to reductions in infiltration and inflow within sewers, due to changes to septage management practices or due to industrial development within the service area.

3.6.2.2 Uncertainty related to data

Measurements contain uncertainty due to both random errors, systematic and gross errors.

Random errors are a consequence of the many uncontrollable and unpredictable errors that exist in the measurement process. They are the effect of many small errors added together. Random errors are indeterminate and can be potentially minimised but never completely removed. They arise in any measurement process and can only be reduced by improving the precision of the measurement.

Systematic errors are non-random errors caused by miss-calibration or malfunction of instruments or the improper location or method for manual or automated sampling. Calibration errors can be reduced through prevention (regular calibration of instrumentation) and partly through data analysis, for example, mass balancing or fault detection (e.g., Lee *et al.*, 2004). Sampling errors can be reduced through better knowledge of the underlying process and increasing of sampling frequency (e.g., Ort *et al.*, 2010). Systematic errors are determinate and can be detected and removed thereby reducing the uncertainty in the measured model inputs. They may be occasional errors or persistent errors.

Gross errors include human oversight and other mistakes while reading, recording, and reading measurements. The most common errors, human errors in the measurement, fall under this category. They can be reduced by the adoption of quality control procedures.

3.6.2.3 Uncertainty related to process modelling

Uncertainty in process modelling arises due to parameter uncertainty (which values to use?), model structure (which model to select?) and errors in implementation.

Parameter uncertainty can be addressed by assigning probability distributions to parameters. In applications where no data are available, a priori uncertainty estimates are obtained from expert knowledge and/or literature. The effects of parameter uncertainty on model outputs can be quantified by the use of Monte Carlo (MC) simulation techniques (Benedetti *et al.*, 2008; Sin *et al.*, 2009).

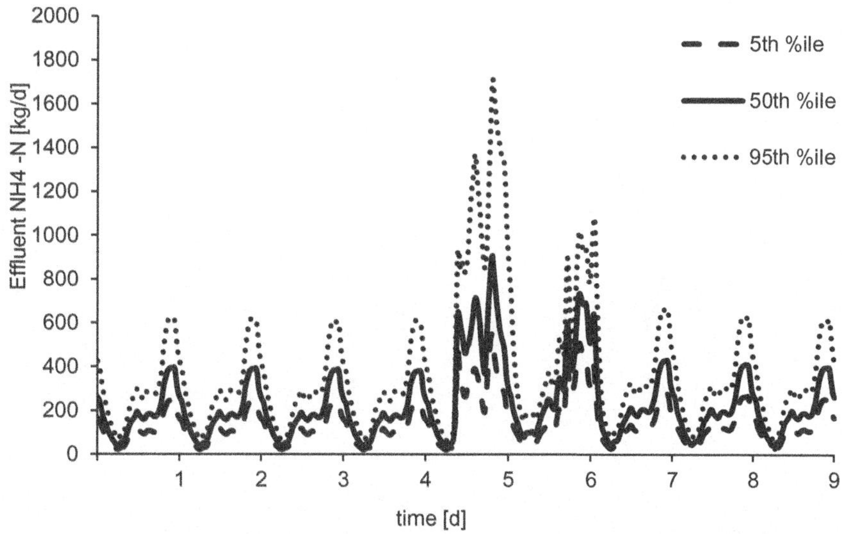

Figure 3.4 Output from an MC simulation with time series for 5th, 50th and 95th percentile values.

Figure 3.4 shows a possible output from such an MC simulation. The plant effluent dynamics are depicted by a range of time series representing percentile values.

In applications where data are available, some parameter values and their uncertainty range can be estimated with the approaches described in Section 3.5.4.

For the practitioner, model structure uncertainty can be addressed in various ways. For example, he/she may want to repeat the modelling exercise with a different model structure or integrate his/her own model structure extensions or reductions (Rieger *et al.*, 2013).

Uncertainty due to modelling errors can be checked by running redundancy checks and elemental balances (e.g., Hauduc *et al.*, 2010). Uncertainty due to software errors can be checked by running a verified model on multiple simulators (reference). Uncertainty due to numerical errors can be captured through the use of multiple simulators, numerical accuracy can be checked by changing the solver properties (such as time step size or solver type and accuracy).

3.6.3 Sensitivity analysis

Typically, a sensitivity analysis is required to prioritise the sources of variability and uncertainty. 'Local' methods analyse how variation in one of the parameters affects the model output while all other parameters are held at the nominal values. 'Global' methods analyse how variation in one parameter affects the model output while all the other parameters are also varying within their uncertainty ranges. Such global methods are quite easy to implement, although some require many simulations (e.g., Benedetti *et al.*, 2011; Neumann, 2012; Sin *et al.*, 2011). An example of such a global analysis is depicted in Figure 3.5.

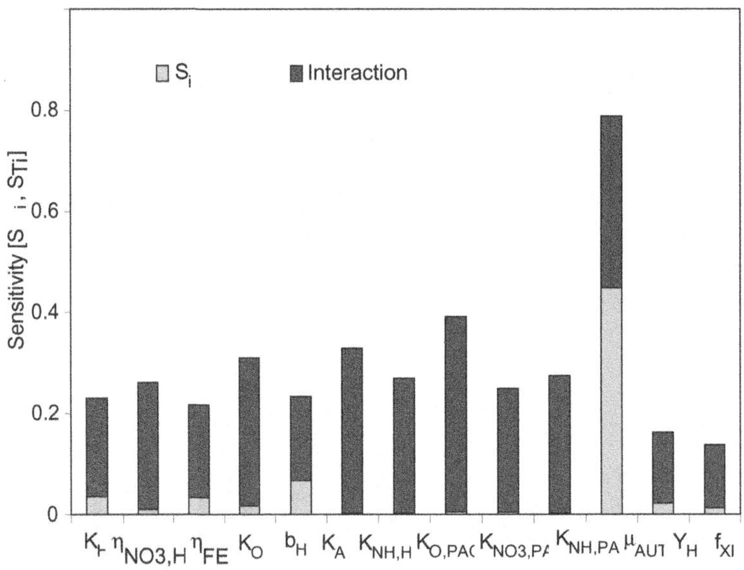

Figure 3.5 Example of a global sensitivity analysis for a membrane bioreactor. Grey section: direct impact of the parameter on the model output uncertainty. Black section: indirect impact of a parameter due to its interaction with all other parameters (Cosenza *et al.*, 2011).

The grey section of the bars quantifies the direct influence of the parameter in determining model output uncertainty (as a fraction of model output variance) and the black section quantifies the interaction effect, which is the indirect influence of a parameter due to its interaction with all the other parameters (Cosenza *et al.*, 2011).

3.7 OPPORTUNITIES OF COMBINING MODELS WITH UNCERTAINTY – EXAMPLE

Figure 3.6 illustrates a typical output of a design example where a mathematical model in combination with uncertainty analysis has been implemented. The x-axis represents concentration and y-axis represents costs (diagonal lines) or probability density.

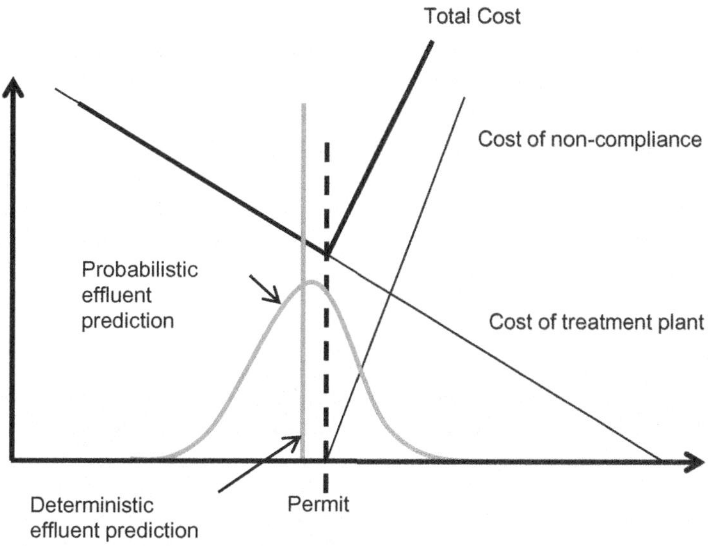

Figure 3.6 Probabilistic design: An optimal design is found by combining probabilistic model predictions with cost functions. X-axis represents concentration and y-axis represents costs (diagonal lines) or probability density in the case of the probabilistic concentration prediction (grey curve). The dashed black vertical line represents the permit limit and the grey vertical line represents the effluent prediction of the deterministic design. Through iteration of the design variables (e.g., tank volume) a design can be found that locates the grey curve in such a way that the expected total cost is minimised.

In this hypothetical example, strict effluent concentration limits (e.g., for maximum month) need to be met. A designer/engineer using a deterministic model will predict a single value for the effluent concentration (vertical grey line in Figure 3.6).

If, however, the engineer expresses her uncertainty about the model parameters with probability distributions and runs an MC simulation, the simulated effluent concentration will become a probability distribution (grey curve in Figure 3.6). Given the permit, the engineer can now either design to a chosen failure probability (e.g., probability of failure = 0.05) or, if cost functions for the treatment plant (capital, operating costs and as non-compliance costs) are available, determine a design that minimises the expected total cost. In Figure 3.6, proposing a smaller design (e.g., smaller tank volumes) would move the entire probability distribution to the right and then the costs due to non-compliance would increase rapidly. Proposing a larger design would move the distribution to the left and then costs would increase due to higher construction- and capital costs. Presumably, the optimal design is the one that leads to a probability distribution that minimises total expected costs, that is, the design with $\int_{-\infty}^{\infty} \{f(predicted\ concentration) \cdot (total\ cost)\} = min$ (see also Reckhow, 1994). This can be seen as an illustrative example of a rational design approach that explicitly deals with uncertainty.

3.8 SUMMARY

Current design approaches still rely heavily on guidelines and on the use of safety factors to account for uncertainty. At the same time, simulators using mechanistic models that capture the details of hydraulic and biochemical dynamics have become common tools in the wastewater engineer's toolbox. If such models are used in design, then uncertainty is typically accounted for in an implicit way, such as designing to stricter standards than those specified. However, these models do also offer the opportunity for explicit considerations of variability and uncertainty. On the one hand, spatial and temporal variability can be examined at higher resolution through CFD and dynamic models, respectively. On the other, (statistical) techniques can be applied that make possible the consideration of measurement uncertainty, parameter uncertainty and model structure uncertainty. This opens up the possibility of moving towards full probabilistic and risk-based designs. At the same time this requires that the limits of predictability are better appreciated by clarifying which real-world phenomena are captured by the models and which are not.

REFERENCES

Alikhani J., Takacs I., Al-Omari A., Murthy S. and Massoudieh A. (2017). Evaluation of the information content of long-term wastewater characteristics data in relation to activated sludge model parameters. *Water Science and Technology*, **75**(6), 1370–1389.

Alvarado A., Vesvikar M., Cisneros J. F., Maere T., Goethals P. and Nopens I. (2013). CFD study to determine the optimal configuration of aerators in a full-scale waste stabilisation pond. *Water Research*, **47**(13), 4528–4537.

Benedetti L., Bixio D., Claeys F. and Vanrolleghem P. A. (2008). Tools to support a model-based methodology for emission/immission and benefit/cost/risk analysis of wastewater systems that considers uncertainty. *Environmental Modelling & Software*, **23**(8), 1082–1091.

Benedetti L., Claeys F., Nopens I. and Vanrolleghem P. A. (2011). Assessing the convergence of LHS Monte Carlo simulations of wastewater treatment models. *Water Science and Technology*, **63**(10), 2219–2224.

Bixio D., Parmentier G., Rousseau D., Verdonck F., Meirlaen J., Vanrolleghem P. A. and Thoeye C. (2002). A quantitative risk analysis tool for design/simulation of wastewater treatment plants. *Water Science and Technology*, **46**(4-5), 301–307.

Bott C. B. and Parker D. S. (2011). Nutrient Management Volume II: Removal Technology Performance & Reliability, WERF Research Project under Nutrient Challenge, WERF Report NUTR1R06k, Water Environment Research Foundation, Alexandria, VA, USA.

Corominas L., Flores-Alsina X., Muschalla D., Neumann M. B. and Vanrolleghem P. A. (2010). Verification of WWTP design guidelines with activated sludge process models. Proceedings of the 83rd Annual Water Environment Federation Technical Exhibition and Conference (WEFTEC 2010), 4–6 October, New Orleans, Louisiana, USA, pp. 137–146.

Cosenza A., Mannina G., Vanrolleghem P. A., Viviani G. and Neumann M. (2011). Global sensitivity analysis in ASM applications: comparison of the SRC and Extended-FAST method for a UCT-MBR model. Proceedings of the 8th IWA Symposium on Systems Analysis and Integrated Assessment, Watermatex, 2011, San Sebastian, Spain, pp. 543–552.

Dominguez D., Worch H., Markard J., Truffer B. and Gujer W. (2009). Closing the capability gap: strategic planning for the infrastructure sector. *California Management Review*, **51**(2), 30–50.

Flyvbjerg B., Bruzelius N. and Rothengatter W. (2003). Megaprojects and Risk: An Anatomy of Ambition. Cambridge University Press., UK, ISBN-13: 978-0521009461.

Gernaey K., Flores-Alsina X., Rosen C., Benedetti L. and Jeppsson U. (2011). Dynamic influent pollutant disturbance scenario generation using a phenomenological modelling approach. *Environmental Modelling & Software*, **26**(11), 1255–1267.

Gresch M., Brügger R., Meyer A. and Gujer W. (2009). Compartmental models for continuous flow reactors derived from CFD simulations. *Environmental Science Technology*, **43**(7), 2381–2387.

Gresch M., Armbruster M., Braun D. and Gujer W. (2011). Effects of aeration patterns on the flow field in wastewater aeration tanks. *Water Research*, **45**(2), 810–818.

Hauduc H., Rieger L., Takács I., Héduit A., Vanrolleghem P. A. and Gillot S. (2010). A systematic approach for model verification: application on seven published activated sludge models. *Water Science and Technology*, **61**(4), 825–839.

Lee J. M., Yoo C. K., Choi S. W., Vanrolleghem P. A. and Lee I. B. (2004). Nonlinear process monitoring using kernel principal component analysis. *Chemical Engineering Science*, **59**(1), 223–234.

Martin C. and Vanrolleghem P. A. (2014). Analysing, completing, and generating influent data for WWTP modelling: a critical review. *Environmental Modelling & Software*, **60**, 188–201.

Mc Cormick J. F., Johnson B. and Turner A. (2007). Analyzing risk in wastewater process design: Using Monte Carlo Simulation to move beyond conventional design methods. In Proceedings WEFTEC2007 pp. 1507–1521, San Diego, California, USA.

Neumann M. B. (2012). Comparison of sensitivity analysis methods for pollutant degradation modelling: a case study from drinking water treatment. *Science of the Total Environment*, **433**, 530–537.

Neumann M. B. and Gujer W. (2008). Underestimation of uncertainty in statistical regression of environmental models: influence of model structure uncertainty. *Environmental Science & Technology*, **42**(11), 4037–4043.

Ort C., Lawrence M. G., Reungoat J. and Mueller J. F. (2010). Sampling for PPCPs in wastewater systems: comparison of different sampling modes and optimisation strategies. *Environmental Science and Technology*, **44**(16), 6289–6296.

Rehman U., Audenaert W., Amerlinck Y., Maere T., Arnaldos M. and Nopens I. (2017). How well-mixed is well mixed? Hydrodynamic-biokinetic model integration in an aerated tank of a full-scale water resource recovery facility. *Water Science & Technology*, **76**(8), 1950–1965.

Reckhow K. H. (1994). Importance of scientific uncertainty in decision-making. *Environmental Management*, **18**(2), 161–166.

Reichert P. (2009). Environmental Systems Analysis (Lecture Notes). Eawag and ETH, Switzerland.

Rieger L., Gillot S., Langergraber G., Ohtsuki T., Shaw A., Takacs I. and Winkler S. (2013). Guidelines for Using Activated Sludge Models, Scientific and Technical Report No. 22, International Water Association (IWA), London, UK.

Rosen C., Rieger L., Jeppsson U. and Vanrolleghem P. A. (2008). Adding realism to simulated sensors and actuators. *Water Science and Technology*, **57**(3), 337–344.

Sharifi S., Murthy S., Takács I. and Massoudieh A. (2014). Probabilistic parameter estimation of activated sludge processes using Markov Chain Monte Carlo. *Water Research*, **50**, 254–266.

Sin G., Gernaey K. V., Neumann M. B., van Loosdrecht M. C. M. and Gujer W. (2009). Uncertainty analysis in WWTP model applications: a critical discussion using an example from design. *Water Research*, **43**(11), 2894–2906.

Sin G., Gernaey K. V., Neumann M. B., van Loosdrecht M. C. M. and Gujer W. (2011). Global sensitivity analysis in wastewater treatment plant model applications: prioritizing sources of uncertainty. *Water Research*, **45**(2), 639–651.

Villez K., Del Giudice D., Neumann M. B. and Rieckermann J. (2020). Accounting for erroneous model structures in biokinetic process models. *Reliability Engineering & System Safety*, **203**, 26.

Chapter 4

Available methods for uncertainty analysis in model-based projects – critical review

4.1 INTRODUCTION

This chapter reviews and summarises the uncertainty analysis methods described in the published literature from the wastewater treatment field. The objective of the review is to capture the breadth of the state of uncertainty analysis within wastewater treatment and does not attempt to focus on a detailed evaluation of individual methods. Most publications reviewed date between 1958 and 2009, with key papers added between 2009 and 2013. Some of the more popular methods referenced in this chapter are illustrated in more detail in Appendix B. Appendix C includes the full list of papers reviewed by the Task Group (including about two topics not covered in this chapter: on-line control and regulatory issues).

Appendix C also includes more recent publications (2011–2019) covering a wide range of uncertainty topics (not reviewed in this chapter).

The presentation in this chapter is of a highly technical nature. Much of the discussion assumes a familiarity with the mathematics that underlie probability theory. The chapter is not meant to be a statistics compendium; it is rather structured as a review article to point interested and theoretically inclined individuals to the body of literature containing examples and discussion of how uncertainty analysis has been applied to wastewater treatment problems. Many practitioners that use treatment plant models for design, or to analyse operational issues, may not find that the material within provides them with guidance that can be applied to their day-to-day work. The connection between this material, and its potential practical applications is the subject of Chapter 5.

Table 4.1 lists the main methods available for uncertainty assessment in model inputs, model parameters, model structure and model-based decision-making. The methods covered in this chapter address **variability and quantifiable uncertainty**. Additional details on the methods referenced in Table 4.1 can be found in Appendix B.

Editor(s): Evangelia Belia, Lorenzo Benedetti, Bruce Johnson, Sudhir Murthy, Marc Neumann, Peter Vanrolleghem and Stefan Weijers
doi: 10.2166/9781780401034_0047

Table 4.1 Uncertainty assessment methods (see Appendix A for definitions).

Category	Method
Model inputs	Summary statistics Statistical tests Outlier detection Data reconciliation Principal components analysis (PCA)
Model parameters	Estimation of inference (confidence) region Bayesian statistics
Model structure	Compartmental modelling
Model-based decision making	Monte Carlo simulation

4.2 METHODS AND LITERATURE REVIEW RESULTS SUMMARY

Literature searches were conducted on ISI/Web of Knowledge (Science), Compendex, Scopus, and Pollution Abstracts and Toxicology Abstracts. The list of wastewater treatment uncertainty references collected is included in Appendix C.

Table 4.2 summarises the results of the search and provides a framework to synthesise the considerable breadth of the topic and size of the literature search results into discrete topics. The categorisation in Table 4.2 is based on subjective judgement and there are several references assigned to one category that address other categories.

It must be noted that not all of the above categories could be covered comprehensively in a succinct manner within this chapter. Specifically, categories 4 and 5 are not covered in this chapter. However, the

Table 4.2 Literature search results.

No.	Category	Description	Number of Papers
1	Input and parameters	References that provide information on the uncertainty in model parameters (single values for steady-state models) or in input time series (inputs for dynamic models)	51
2	Model structure	References that address uncertainty generated from the structure of a wastewater model or references that address mathematical concepts related to uncertainty within the context of a wastewater treatment model	58
3	Propagation of uncertainty	References that address uncertainty evaluation of one or more different treatment trains or plant-wide alternatives in model-based decision-making	119
4	On-line control signals and strategies	References that consider the uncertainty of an on-line measurement or references dealing with the use of on-line signals within a real-time control loop	92
5	Fate of pollutants in the environment	References that address the uncertainty associated with the presence of pollutants in the environment and resulting regulatory (WRRF effluent standards) issues	85

reader may review the references under these categories in Appendix C as a starting point for further information on these subjects.

This chapter focuses on the first three categories listed in Table 4.2. Within each category, abstracts were further screened, and a number of papers was selected for detailed review.

4.3 ASSESSMENT OF INPUT AND PARAMETER UNCERTAINTY

4.3.1 Input uncertainty (measurement errors)

This area of research focuses on quantifying the uncertainty in model inputs and on the use of techniques to minimise the model input uncertainty before performing further analysis. Input uncertainty is due to measurement errors and here the methods implemented in the literature to quantify them are discussed. Measurements contain uncertainty due to random, systematic and gross errors (for a definition of measurement errors, see Chapter 3, Section 3.6.2.2 and Appendix A).

Methods for quantifying the uncertainty stemming from the assumption that the measured data are an unbiased estimate of the underlying population have not been included in this review.

4.3.1.1 Overview of statistical techniques used in measurement error detection

The topics addressed in the reviewed papers include quantifying the uncertainty in data and data collection methods and identifying systematic errors in data. The uncertainty in measured data is typically assessed using statistical techniques. Standard Methods for the Examination of Water and Wastewater (1998) provides background on measurement uncertainty and basic statistical techniques used to quantify this uncertainty as applied to the examination of wastewater. Skoog *et al.* (1995) describe the use of these uncertainty evaluation techniques in the field of analytical chemistry.

The precision of a measurement can be determined by replicate measurements and by the calculation of the standard deviation and variance of the replicates. A common technique for assessing the accuracy of measurements and detecting systematic errors in an analytical method or an instrument is to analyse a sample whose composition is accurately known (i.e., a calibration check standard). Statistical hypothesis testing is then used to assess whether the difference between the measured value and the known calibration check standard value could be caused by random error or systematic error. Outlier detection tests are available for detection of large biases but must be used cautiously (Skoog *et al.*, 1995). More advanced techniques, such as data reconciliation, may be more suitable for outlier detection where applicable (see discussion below).

4.3.1.2 Error propagation

In cases where results are computed from multiple sources of experimental data or a calibration curve is used to provide the measured value, it is necessary to determine how the error in the measured values is transmitted to the results. For general non-linear functions, a few uncertainty assessment approaches are available such as the law of propagation of uncertainty, Monte Carlo simulation, and empirical sensitivity studies based on designed experiments (Standard Methods for the Examination of Water and Wastewater, 1998). A common approach is to use the law of propagation of uncertainty where the function of interest (e.g., the calibration equation) is linearised using a first-order Taylor-series expansion about the key variables and then the variance formula for a linear sum of variables is used to calculate the total variance (Box *et al.*, 1978).

In the wastewater treatment literature, these statistical techniques have been used to quantify the uncertainty in various types of wastewater measurements. Friedler and Butler (1996) used basic statistical techniques to quantify the inherent uncertainty in the quantity and quality of wastewater discharged from domestic appliances. The analysis was based on data from two surveys conducted in the United Kingdom. The appliance volumes, pollutant loads, and frequency of use were not found to be normally distributed variables. The authors suggest that Monte Carlo analysis could be used to help quantify the combined effects of the uncertainties at the household level on the overall uncertainty in the wastewater flow rate and concentrations within a sewer system.

4.3.1.3 Examples of measurement error detection

Joannis *et al.* (2008) studied the uncertainty in wastewater turbidity measurements. They found that the major sources of uncertainty were in the standard solutions used for calibration and the nonlinearity of the calibration curve. Bertrand-Krajewski *et al.* (2007) compared the uncertainties in COD measurements between standard laboratory techniques, small tube tests (STT) employing a photometer, and field UV–visible spectrometry. They found that standard laboratory methods and small tube tests had similar levels of uncertainty but had different mean values, indicating that specific calibration functions are needed to correct systematic errors if high accuracy is required or the methods are compared. Bertrand-Krajewski *et al.* (2007) found that the use of low-frequency sampling is the major source of uncertainty with standard laboratory methods and methods for COD determination. UV–visible spectrometry was found to have a similar level of uncertainty as standard laboratory methods but only under strictly controlled conditions.

Rieger *et al.* (2005) evaluated the uncertainty of on-line sensors at WRRFs using comparisons between independent measurements of the same sample (i.e., sensor and a reference laboratory method). The comparison is based on a linear regression fitted between the sensor and reference measurements. The authors assess whether the linear regression is applicable by considering a relationship between the variables (using an *F*-test), checking the linearity between the sensor and reference measurements (using an *F*-test), checking for outliers, and checking the homogeneity of the variances. If the linear regression is applicable, statistical tests on the regression predictions and the regression slope and intercept are used to assess whether the regression equation is significantly different from the perfect correlation (i.e., slope = 1 and intercept = 0), indicating the presence of systematic errors. If no systematic errors are detected, the total uncertainty is represented by a confidence interval for the regression predictions, assuming a perfect correlation between the sensor and reference measurements. If systematic errors are detected, the bias is quantified by the linear regression fit between the sensor and reference measurements. The random errors are quantified by the confidence interval for the regression predictions.

4.3.1.4 Multivariate statistical methods

Robinson *et al.* (2005) used multivariate statistical methods to identify outliers in water quality data. They advocated the use of multivariate statistical methods due to the correlation between plant variables. Methods discussed include: Mahalanobis distance, jack-knife distance, and Hadi's method. These methods are more applicable than univariate methods but do not account for the serial correlation of the variables over time. Multivariate statistical control methods are available that can address serial correlation, as discussed below.

There are also more advanced statistical methods available for detecting and removing systematic errors and gross errors. These methods range from statistical process control and fault detection methods to data reconciliation.

4.3.1.5 Statistical process control and fault detection methods

Statistical process control techniques involve monitoring process variables over time using statistical control charts. The variables of interest are charted over time and compared to control limits to determine if the process is within control (i.e., its correlation structure is unchanged). The methodology is used to distinguish between common cause variability and special causes. Typically, these methods are used to study process variability but can also be used in the current context to detect sensor or measurement process faults leading to large systematic or gross errors.

Because WRRF measurements can exhibit autocorrelation, seasonality, and non-constant variance (Berthouex, 1989), it can be difficult to apply traditional control charts such as Shewhart or cumulative sum (CUSUM) charts to the measured process variables themselves. One option discussed by Thomann *et al.* (2002) and Thomann (2008) is to create a control chart that tracks the difference between the sensor values and reference values at a WRRF. Unfortunately, this approach may not always be practical if reference measurements are not taken at a suitable interval. Another approach is to fit a time-series model such as an auto-regressive integrated moving average (ARIMA) model to normal operating data and then use the model as a charting tool (Berthouex, 1989). The model would be used to continually predict process data given the previous data and the difference between these predictions and the actual measurements (i.e., residuals), are plotted on a conventional control chart. When the measurements are collected normally, the residuals will be independent, random, and have constant variance.

A simpler alternative is to construct exponentially weighted moving average (EWMA) charts. The one-step-ahead prediction errors (i.e., residuals between predictions and actual measurements), calculated using the EWMA statistic, can be plotted on a traditional control chart. As discussed by Montgomery and Mastrangelo (1991), the EWMA approach can be a reasonable approximation of the ARIMA model approach in many cases. For suitably selected value of the EWMA filter constant, the EWMA statistic is an excellent one-step-ahead predictor for processes where the mean does not shift too rapidly, and the observations are positively auto-correlated.

The limitation of univariate control charts is that they do not consider the correlation between the variables within the process. Some researchers have looked at the use of multivariate statistical techniques such as principal components analysis (PCA) and partial least squares (PLS) to monitor process data. PCA involves projecting multivariate data into a lower dimensional or latent variable space. The variables in the lower dimensional space are uncorrelated and explain the majority of the variance in the data. PLS is a latent variable regression method used when multivariate input and/or output data are available. The PLS regression model captures the correlation between the inputs and outputs in a lower dimensional variable space. PCA and PLS models used for monitoring are built using data from normal operation so that they model the normal measurement variability and correlation.

Using a PCA model for example, one can create control charts for outputs from the model including t-scores, the Hotelling T^2 statistic, and the squared prediction error (SPE) to detect unusual measurements in a multivariate context (Kourti & MacGregor, 1995). Although PCA and PLS consider static covariance relationships, they can be adapted to the analysis of dynamic data by including time-lagged data into the data matrices so that the correlation over time is included into the models. In a wastewater treatment context, some examples of latent variable monitoring are discussed by Lennox and Rosen (2002) and Tomita *et al.* (2002).

4.3.1.6 Data reconciliation

Data reconciliation is a technique used to adjust process measurements so that they are consistent with known conservation laws and other process constraints. The procedure requires a set of redundant measurements to verify that the conservation laws have been obeyed. Optimal data reconciliation is a constrained least-squares problem requiring the minimisation of a weighted sum of the measurement adjustments subject to the process constraints. The weighting matrix is typically the inverse of the variance–covariance matrix of the errors in the measurements. The weighting may be selected based on previous experience, calculated using the sample variance for the data, or using robust estimators (Chen *et al.*, 1997). The measurements in the data reconciliation procedure are weighted inversely to their variance so that measurements with large variance are adjusted more than those with a smaller variance. Therefore, the success of the method is dependent on the use of reasonable variance estimates.

In the context of biochemical reactions, data reconciliation has been studied by Van der Heidjen *et al.* (1994a, b, c). Mass balancing has been discussed in a WRRF context by Nowak *et al.* (1999) and Barker and Dold (1995), while more formal data reconciliation analyses have been reported by Meijer *et al.* (2002), Puig *et al.* (2008), and Thomann (2008). Recently, Rieger *et al.* (2010) discussed data reconciliation for WRRF simulation studies. Their focus was on planning measurement campaigns so that high-quality data can be collected. They discuss the use of basic reliability checks and manual checking of mass balances to verify the quality of the data and to identify systematic errors.

In a WRRF context, it is most common to reconcile flow and total phosphorus measurements across the plant, and suspended solids measurements around clarifiers and thickening and dewatering processes. COD and total nitrogen measurements could be also potentially reconciled using mass balances, but this typically requires measurements not typically collected such as the oxygen utilisation rate (OUR), oxygen transfer parameters, and nitrogen gas flows.

Data reconciliation can be performed using either a steady-state or dynamic analysis. Steady-state data reconciliation is commonly performed using averaged measurements over a period of approximate steady-state or zero accumulation. Examples in a WRRF context are provided by Meijer *et al.* (2002) and Puig *et al.* (2008). Puig *et al.* (2008) suggest that the data be averaged over a period of at least two to three sludge retention times. In the case of steady-state reconciliation, the process constraints (typically mass balances) are assumed to be known and the measurements are considered to be stochastic.

In dynamic data reconciliation, the process constraints are typically dynamic process models so that uncertainty is considered to exist in the model structure and parameters, and the measured data. Dynamic data reconciliation can be conveniently performed in a simulation environment by minimising a weighted least-squares function of the measurement adjustments, subject to the WRRF model, over successive time horizons or windows. This method is known as the horizon method (Romagnoli & Sanchez, 2000). In the horizon method, the initial values of the model states for each time horizon are the optimisation variables.

Dynamic data reconciliation can also be performed using a filtering approach based on the extended Kalman filter (Romagnoli & Sanchez, 2000). In this context, the filter acts as a state estimator which takes the states predicted by the model and adds the filtered difference between the measured and predicted model outputs. The filtering approach has an advantage in that its calculations are recursive and do not require iteration as in the horizon method. The horizon method is thought to be better suited to slower processes (Cameron *et al.*, 1992), such as biological growth, while the filtering method is thought to be better suited to faster processes.

Following data reconciliation, gross error detection techniques can be used to identify and eliminate systematic errors caused by sensors and other faults. Gross error tests involve statistical hypothesis testing on the least-square's objective function (which is a Chi-square variable), and on the ratios of the measurement adjustments and the mass balances errors to their standard deviations. Meijer *et al.* (2002) and Thomann (2008) illustrate the use of simple gross error detection techniques in a WRRF context. More sophisticated techniques are discussed by Crowe (1996) and involve the use of tests of maximum power in detecting gross errors for the case of a single gross error and the use of PCA in cases where multiple gross errors exist.

The use of formal data reconciliation in the wastewater treatment field has been limited due to a lack of data, the complexity of the solution procedure, and the lack of availability of software dedicated to the solution of the data reconciliation problem. For simulation studies, most model calibration protocols typically recommend basic reliability checks and manual evaluation of mass balances to verify the quality of the data and to identify systematic errors (Rieger *et al.*, 2013). This is expected to change in the future as online instrumentation becomes common in WRRFs, and modelling and other software vendors add the necessary tools to their products. Data may be taken at different intervals, contain missing measurements, have redundancy, and could sometimes be erroneous. Adaptation of existing mathematical tools into the databases and SCADA systems of full-scale plants will be required in order to promote the progressive incorporation of advanced monitoring systems, decision support systems, and plant-wide controllers.

4.3.2 Parameter uncertainty

Whole plant models are complex and have hundreds of parameters, all with some uncertainty. While most of those parameters can be assumed to be fixed, others require to be considered uncertain, given their importance towards the results of the model use.

Uncertainty in model parameters arises from many sources such as the model structure, their measurement error (in case they are directly or indirectly measured), the choice of experimental conditions used for model calibration, the calibration data, and the objective function or criterion used for parameter estimation.

4.3.2.1 Inference vs. confidence regions

Uncertainties in model parameters are typically assessed during the process of parameter estimation. See Bard (1974), Draper and Smith (1989), and Bates and Watts (1988) for the theory of nonlinear parameter estimation. Parameter estimation problems are often solved using maximum likelihood estimation. Depending on the assumptions on the error structure, the number of response variables, and the available information on the variance and covariance of the errors, the objective function minimised during the procedure ranges from ordinary least squares, to weighted least squares, to the Box–Draper criterion (Box & Draper, 1965).

Parameter uncertainty is typically assessed following parameter estimation, through the calculation of approximate joint confidence regions for the parameters and approximate confidence limits on individual parameters. The inference regions and limits or bands are often estimated by extending linear regression theory. The model residuals are linearised using a Taylor-series expansion and analogous formulas as those used for linear regression inference regions and bands are developed (Draper & Smith, 1989).

The inference region formulas require the calculation of the variance–covariance matrix for the parameters. The variance–covariance matrix is often approximated as the inverse of the Hessian matrix of the objective function (i.e., matrix of second derivatives of the objective function) multiplied by a scale factor at the solution to the parameter estimation problem (Bard, 1974). In the linearisation approach, the Hessian matrix is calculated using first-order model sensitivity coefficients only (Gauss–Newton approximation). First-order model sensitivity coefficients, which express the local sensitivity of the process model to infinitesimally small changes in the model parameters, are defined as the partial derivatives of the model with respect to the model parameters. The sensitivity coefficients can be determined using finite-difference approximations, by solving the model sensitivity equation (Leis &

Kramer, 1988), using variational methods, or by automatic differentiation (De Pauw & Vanrolleghem, 2003). Alternatively, it is also possible to approximate the variance–covariance matrix using the full Hessian matrix which requires the calculation of second-order sensitivity coefficients which can be calculated as shown by Guay and Maclean (1995).

4.3.2.2 Application to wastewater treatment models

In the context of biokinetic models of the activated sludge process, the calculation of approximate inference regions for the model parameters has been discussed by numerous researchers including Vanrolleghem *et al.* (1995), Vanrolleghem and Keesman (1996), Brouwer *et al.* (1998), Petersen (2000), Petersen *et al.* (2000), Dochain and Vanrolleghem (2001), Marsili-Libelli and Tabani (2002), Sin (2004), Checchi and Marsili-Libelli (2005), and De Pauw (2005). Parameter estimation techniques are most often applied in a wastewater treatment context when fitting biokinetic models to respirometric batch experiments. Formal parameter estimation (e.g., using maximum likelihood estimation) is not recommended for calibrating entire WRRF models to historical plant data due to the lack of data, the complexity of the models, and the correlation between the model parameters (Petersen, 2000; Vanrolleghem *et al.*, 2003). Historical plant data are rarely suitable for estimating complex model parameters and for assessing their uncertainty due to missing data, inconsistencies in the data, limitations in the ranges of the variables due to process control, confounding effects between variables, and variations in unmeasured variables (Box *et al.*, 1978). Typically, model calibration focuses on influent characterisation, accurate modelling of plant hydraulics and aeration, and manual adjustment of some model parameters to achieve a reasonable fit between the measured data and model outputs.

In the context of parameter estimation, it is possible to design experiments that minimise the uncertainty in the estimated parameters. One common approach, introduced by Box and Lucas (1959), is to minimise the volume of the parameter confidence region. This involves minimising the determinant of the inverse of the variance–covariance matrix. A sequential strategy is often used, as the best set of experimental conditions depends on the parameter values (Box *et al.*, 1978). Vanrolleghem *et al.* (1995) discuss the use of optimal experimental design in the context of activated sludge models (ASMs) and list alternative optimal design criteria.

The main drawbacks of these approximate parameter uncertainty assessment methods are that they assume that only the response variables in the parameter estimation procedure contain uncertainties, they use an approximation to the variance–covariance matrix, and they are specific to the local solution to the parameter estimation problem.

4.3.2.3 More sophisticated methods

It is possible to reformulate the parameter estimation problem using an error-in-variables (EIV) approach so that both the independent and dependent variables (i.e., all model inputs) are considered to contain uncertainties (Bard, 1974; Romagnoli & Sanchez, 2000). This becomes a simultaneous data reconciliation and parameter estimation problem.

Better estimates of the parameter uncertainties can be obtained using the Monte Carlo method where the parameter estimation problem is solved repeatedly for different simulated samples of the measured model inputs leading to a distribution of parameter estimates (Bard, 1974). Another option is to use the technique known as profiling (Bates & Watts, 1988), after parameter estimation, to obtain exact marginal likelihood intervals for the model parameters. Cox (2004) used Bayesian statistics to develop uncertainty distributions for the parameters in the ASM1 model. That procedure involves combining expert opinion (prior distribution) and measured or calibrated parameter values into a single posterior distribution known as a universal distribution. The method is promising but the specific application given by Cox may not be useful given that many of the calibrated values are taken from subjective calibration studies involving historical plant data.

4.4 ASSESSMENT OF MODEL STRUCTURE UNCERTAINTY

A portion of the references identified by the literature search addressed the fundamental wastewater process model uncertainty issues of (1) model structure and (2) mathematical methods. These areas of the literature search may be of most interest to researchers investigating fundamental information and approaches to uncertainty assessment in wastewater treatment engineering. Some of the most compelling works published in these areas are discussed in the following sections.

4.4.1 Macroscopic vs. microscopic mixing scales

Several researchers have investigated structural issues at the core of activated sludge models (ASMs) that arise from the conceptual basis of some state variables and assumptions used in modelling of the completely stirred tank reactor (CSTR) configuration. While Danckwert's (1958) and Zwietering's (1959) seminal publications on residence time distribution and reactor modelling identified the influence of the nature of the reactants, reaction rates, and local mixing scale on CSTR reactor analysis, and while chemical engineering text books (e.g., Levenspiel, 1999; Rawlings & Ekerdt, 2002) have further formalised the concepts to recognised limiting cases of 'complete segregation' and 'maximum mixedness', it was perhaps not until Gujer (2002) that the importance of these concepts to ASMs was noted.

Gujer (2002) observed that ASM state variables representing cell internal storage products are conceptually linked to a local environment (an individual cell), and that reaction rates in ASMs are not necessary first order. There are, therefore, resulting consequences of applying ASMs on a macroscopic and microscopic mixing scales that impact the applicability of kinetic parameters determined for and by different reactor configurations. Gujer (2002) considered a model with a simple single substrate with cell storage product and applied it in both the typical macroscopic fashion and in a microscopic fashion which tracked individual bacteria and used a probabilistic rule to control the residence time of a bacterium within a CSTR zone. While Gujer (2002) notes that this simple model is not directly applicable to any relevant system, the results he presents identify the nature of this basic model structure issue and lead to important conclusions regarding the differences between the determination and applicability of kinetic parameters for sequencing batch reactors and flow through systems.

Gujer (2002) suggests that (1) there may be an inherent amount of quantifiable uncertainty in ASM results associated with reactants' local environment and residence time and (2) that there may be uncertainty or inaccuracy induced into modelling efforts by application of kinetic parameters estimated from differing flow schemes.

Schuler (2005, 2006) extended the concepts of Gujer (2002) to a model that included the competition between phosphorus accumulating organisms and heterotrophic organisms, without nitrifiers and the potential interference of nitrate, and applied the model to reactor configurations relevant to activated sludge treatment systems. Schuler (2005, 2006) illustrated the differences in results that occur between a lumped parameter (macroscopic) model structure and a (microscopic) model structure that includes the distributed state of reactant residence times and concluded that the lumped parameter approach consistently predicted better effluent phosphorus performance.

Curlin *et al.* (2004) also reported on this issue and applied formal concepts from the field of chemical engineering. They used activated sludge model (ASM) No. 1 for a laboratory-scale membrane bioreactor. They established the macroscopic mixing characteristics of their system through tracer studies and arrived at a CSTR combination that fits the experimental tracer results. They then solved the ASM not only using the CSTR assumption typically employed (the ideal mixing case, an assumption of constant reactant concentration over the reactor volume) but also microscopic mixing scale limiting cases of complete segregation and maximum mixing.

Curlin *et al.* (2004) conclude that (1) in ASMs that do not include cell internal storage product state variables, the non-first-order reaction rates may result in un-quantified model structure uncertainty if only the ideal mixing CSTR assumption is considered, (2) the model structural uncertainty generated by imperfect knowledge of microscopic mixing might be quantified through consideration of the limiting cases and (3) the magnitude of the uncertainty may be significant. A validation of the simplified transport models is suggested, compared to more sophisticated approaches.

It is worthwhile to note that other researchers, including Lee *et al.* (1999a, b) and Makinia and Wells (2000a, b) have considered the impact of mixing conditions and residence time distribution in ASMs, using an advection and dispersion equation approach. This line of development, as well as combined unit process modelling and computation fluid dynamics modelling, may also be useful for the identification or reduction of non-ideal mixing contributions to model structure uncertainty.

4.4.2 Unquantified model structure uncertainty

Other researchers have reported on a variety of specific structural issues related to the modelling of wastewater treatment processes that may contribute to what is, at this time, un-quantified model structure uncertainty. Abusam and Keesman (2002) carried out a factorial sensitivity analysis on the use of the double-exponential function in secondary clarifier models and concluded that the model had a structural problem related to the prediction of solids in the underflow stream. Haider *et al.* (2003) reported on experimental results that illustrated that the characterisation of modelled influent non-biodegradable substrate was not independent from the system sludge age and concluded that two such influent biodegradable state variables may be required for models of short sludge age systems. Lavallee *et al.* (2005) similarly noted that observed kinetic parameters depend on substrate, process configuration and sludge, and introduce an ASM within the ASM framework that mimics enzyme induction and may lead towards models applicable over more widely varying conditions. Sin and Vanrolleghem (2006) observed that, even with constant influent conditions, the ASM2d model structure had to be adapted in response to changes in system behaviour observed for three different operational scenarios to match experimental findings. Neumann and Gujer (2008) provided an analysis of model structure uncertainty by generating synthetic data with one model structure (using Tessier rate equation) and fitting it with another putative model (using the Monod rate equation). They illustrated the application of a range of methods for analysing model fit and for propagation of parameter uncertainty to modelling results. This made possible a comparison of model predictions with parameter uncertainty addressed to be compared to the 'true' result and illustrated that the propagation of parameter uncertainty was not adequate to address an error in model structure. Neumann and Gujer (2008) concluded that uncertainty estimates obtained from regression of time-continuous environmental systems should be used with caution.

This is a small sampling of reports in the literature that may be taken to illustrate the degree of structural uncertainty present in wastewater treatment models. In general, these reports suggest recognition that the current wastewater models may be calibrated only to narrow and specific ranges of influent, operating and process configurations conditions, and that any extension of the use of a model outside its specific range of calibration may induce what is, at this time, an un-quantified degree of uncertainty.

4.4.3 Mathematical methods for quantification of model structure uncertainty

While some researchers have focused on and provided useful information on issues of uncertainty within wastewater treatment model structure, other researchers have considered issues related to the mathematics and numerical methods of the quantification of uncertainty.

4.4.3.1 Monod growth model

Tenno and Uronen (1995) applied an ASM1-like model structure within a stochastic model to arrive at a method of carbon removal process control using on-line instrumentation. Kops and Vanrolleghem (1996) investigated the incorporation of uncertainty analysis into wastewater modelling predictions through consideration of the Monod growth model. They compared three methods for approximating prediction uncertainty: Monte Carlo simulation, Monte Carlo simulation with stochastic parameters, and stochastic differential equations. They identified the Monte Carlo simulation as having the disadvantage that for a dynamic simulation all the parameters must stay constant during one model run. The other two alternatives permit time-varying parameters within one model run. They identified the disadvantages of the stochastic differential equation alternative as requiring parameters to have Gaussian distributions and that stable solutions may be limited to a certain range of parameter values. The Monte Carlo simulation with stochastic parameters did not have the disadvantages of the stochastic differential equation alternative. For the situation examined, Kops and Vanrolleghem (1996) found the stochastic differential equation alternative to give a higher predicted variable variance than either of the Monte Carlo-based methods. In their case, the stochastic differential equation method generated a variance of almost 40% while the Monte Carlo with stochastic parameters simulation resulted in a variance of 1%. They concluded that the comparison of these two alternatives becomes a question of which result is more realistic. This appears to remain a crucial, valid and unresolved issue.

Omlin and Reichert (1999) provide a comparison of parameter estimation methods and their related model prediction uncertainty for a simple Monod equation model. They concluded that classical frequentist (i.e., least squares) technique is superior in the case of identifiable model parameters but in the case of poor parameter identifiably, a Bayesian approach is recommendable. Rauh *et al.* (2004, 2007) and Krasnochtanova *et al.* (2009) discuss incorporation of parameter uncertainty at the numerical simulation time step interval level. They note that the efficiency of Monte Carlo methods decreases significantly for higher dimensional systems and present numerical solution algorithms that generate upper and lower bounds on variable uncertainty at each step interval.

4.4.3.2 Non-linear dynamical and chaotic behaviour

The final area of the literature review addressed in this subsection is published work which has considered the potential role of non-linear dynamical and chaotic behaviour in the variability and randomness of wastewater treatment process observations and process models. Graham *et al.* (2007) report the experimental demonstration of chaotic instability in biological nitrification. They operated three highly controlled aerobic chemostats. They indicate that their experimental results and analysis suggest broad chaotic behaviour and conclude that nitrification is prone to chaotic behaviour because of a fragile ammonia oxidising bacteria and nitrite oxidising bacteria mutualism. Zhang and Henson (2001) demonstrated the possibility of multiple steady-state solutions for several continuous biochemical reactor models and advocate the use of bifurcation analysis to aid in obtaining more efficient and complete characterisation of model behaviour. Saikaly and Oerther (2004) and Stroot *et al.* (2005) investigated the potential dynamical nature of competition between several species for several resources within activated sludge treatment systems. They report on modelling results as well as fluorescence in-situ hybridisation determination of biomass composition within the pilot reactors. They found the potential for population oscillations within the bacterial community under some operating conditions and conclude that such a dynamical nature could contribute to system stability while confounding the tracking of activated sludge system composition when done by limited grab sampling. Ibrahim *et al.* (2008) present results of static and dynamic bifurcation investigations of an activated sludge system model using a rate equation that includes an inhibition term. Their analyses show a complex variety of dynamic results when inhibition is significant, including periodic attractors, point attractors, and chaotic attractors for realistic ranges of parameter values.

Thus, several publications suggest that the non-linearity and dynamical nature of wastewater treatment process models and, perhaps, the systems themselves result in inherent randomness, periodicity or chaotic behaviour. The tools of this branch of mathematics may, therefore, be useful in characterising uncertainty for some conditions and model applications.

4.5 PROPAGATION OF UNCERTAINTY FOR MODEL-BASED DECISIONS

4.5.1 Review of uncertainty propagation methods

Models are used in process engineering to configure and size facilities to reliably meet effluent quality requirements. The research reviewed for this effort included a variety of biokinetic models used for process engineering including ASM1, ASM2, and ASM3 models as well as 2-D clarifier modelling. Amongst all consulted literature sources, there was a common understanding that the models are both relatively complex and the inputs used for modelling include significant uncertainty. With this basis, the researcher's general goals were then to understand the sensitivity of model results to uncertainty in model inputs, and the evaluation of methods for generating robust designs.

The literature reviewed focused on three main subject areas: model calibration, sensitivity analysis, and design optimisation.

The review was structured following a number of criteria that involve the purpose, method, accuracy, difficulty/simplicity, time to do the analysis, data requirement, applicability and stakeholder interest (e. g., researchers, control decision support, design, operation). The review is summarised in Table 4.3 and in the following sections.

In addressing the input uncertainty for modelling, the majority of the researchers relied on a Monte Carlo approach both for determining uncertainty in model outputs and for calibrating models. Variations on the Monte Carlo approach included:

- Using Spearman's rank correlation to determine model sensitivity to input data (Griborio *et al.*, 2007);
- Applying the Hooke–Jeeves direct search technique for design optimisation (Tansel, 1999);
- Applying various statistical analyses techniques with Monte Carlo outputs for model calibration.

In addition to Monte Carlo techniques, researchers used genetic algorithms and the 'ϵ constraint method' to determine optimal designs. Although these methods appear to have their strengths, the special expertise needed to apply them may not make them useful for most process engineers.

The information provided did not allow for a detailed assessment of the time required for each method. Furthermore, because the methods were applied to several different models with varying levels of complexity, a direct comparison is not possible.

Table 4.3 Review of uncertainty propagation methods and analysis works applied for model-based evaluation.

MODEL CALIBRATION							
Purpose	**Method**	**Accuracy**	**Difficulty**	**Time**	**Data**	**Engineering Task**	**Reference**
Model calibration of ASM3	Monte Carlo (MC)	± 20% for denitrification with sludge digestion	Medium	Not quantified		Estimation of N removal and sludge production	Koch *et al.* (2001)
Model Calibration of ASM2d	MC + results analysis with MAE, RMSE and Janus coefficient		Medium	2 weeks for 500 MC runs on Pentium IV 3 GHz	Plant data (assumed uniformly distributed), uncorrelated kinetic parameters	Calibration of dynamic models	Sin *et al.* (2008)
Estimate enclosures of state variables in a simple ASM model	MC		More difficult	Not quantified, but would be significant for complex model			Kletting *et al.* (2007)
SENSITIVITY ANALYSIS							
Purpose	**Method**	**Accuracy**	**Difficulty**	**Time**	**Data**	**Engineering Task**	**Reference**
Sensitivity analysis for control strategies	MC + input variables classification		More difficult	Not quantified, but MC × 10 Not completely automated			Von Sperling (1993)
Uncertainty and sensitivity with ASM1	MC + Spearman's rank correlation		Medium		Only kinetic variables, not influent characteristics		Huo *et al.* (2004)
Estimate secondary clarifier performance with 2Dc	MC		Relatively simple statistical approach	Uncertain	Floc and settling parameters	CFD modelling of SC and some info on SVI and settling relationships	Griborio *et al.* (2007) Kletting *et al.* (2007)

(*Continued*)

Table 4.3 Review of uncertainty propagation methods and analysis works applied for model-based evaluation (*Continued*).

DESIGN OPTIMISATION							
Purpose	**Method**	**Accuracy**	**Difficulty**	**Time**	**Data**	**Engineering Task**	**Reference**
Dependability of design using non-ASM model	MC + Hooke-Jeeves optimisation	Lower (less sophisticated model)	Simple	Not quantified			Tansel (1999)
Optimised design using a precursor to ASM	Genetic algorithm also briefly discusses coupling with MC	Better probability than non-linear programming approach	High	Not quantified but faster than non-linear programming	Varied process sizing – did not look at variability in kinetic parameters or influent	May be applicable to optimising or calibrating	Doby *et al.* (2002)
Risk-based design (ASM1)	MC	Good (comparing model results to ammonia data)			Plant data and kinetic variables from Cox (2004)		Huo *et al.* (2006)
Low-cost design using ASM3 for BNR	ε constraint method to generate Pareto optimality		More difficult	Not quantified	ASM3/ASM2d standard		Afonso and da Conceição Cunha (2007)
Predict failures of wastewater pipe systems (utility)	Generalised likelihood uncertainty estimation (GLUE)	Sensitive to errors in model and data (subjective)	High (GLUE algorithm more complex than MC)	Not quantified but expected to be long (GLUE requires thousands of model evaluations)	Past failure rate of pipes Hydrological/climate/geological data of the area flows	Operators, control decision support	Franks (1999)
Compute extreme event statistics in the water quality field (e.g., after pollutant load discharge from CSO to lake) (utility)	MC first order reliability model (FORM) + importance sampling (IS) or LHS Random directional sampling (RDS)	FORM/LHS most accurate	High (FORM algorithms quite complex)	Thousands of simulations (e.g., 15 × 2500 for some methods)	Rainfall COD Water quality model Extreme events Input uncertainty characterisation	Researchers	Portielje *et al.* (2000)

Table 4.3 Review of uncertainty propagation methods and analysis works applied for model-based evaluation (*Continued*).

DESIGN OPTIMISATION

Purpose	Method	Accuracy	Difficulty	Time	Data	Engineering Task	Reference
Rank stormwater control strategies under uncertainty	MC ranking methods (mean + sd) uniform distributions	Sensitive to type of ranking method used Depends on the scenario	Medium (MC engineering standard)	500 MC simulations (NO CPU time reported)	Flows, BOD, DO, temperature Hydraulic/biochemical data Control strategies (seven total)	Decision makers	Duchesne *et al.* (2001)
Quantify uncertainty for WRRF design / retrofit	MC	Relative1 measure	Medium	Long (days), depends on computational power and case	Plant data (influent load/composition, size, layout, …)	General applicability (researchers, designers, operators, control)	Rousseau *et al.* (2001)
Uncertainty in estimating the cost of WRRF constructions	Linear regression Fuzzy linear regression Fuzzy goal regression	Sensitive to database used for building regression models	Medium (regression is standard practice)	3 simulations	Database on construction cost of domestic/industrial plants in Taiwan	Decision makes	Chen and Chang (2002)
Risk-based WRRF design (replace safety factors)	MC (method of Rousseau *et al.*, 2001)	Relative1 measure for decision-making	Medium (see data requirement)	Long (days), depends on computational power and the plant in question	Influent load/composition Rainfall Plant model Temperature	Researchers and designers (highly relevant) Dedicated software to generate MC samples + run them	Bixio *et al.* (2002)
Integrated process design and control via global optimisation	Non-linear programming (NLP) Mixed integer optimal control problem (MIOPC) Global optimisation methods	Depends on initial layout and starting point for optimisation	Complicated (NLP programming tedious)	1000 CPU seconds (1 s is worth of the computer's processing time) (depends on different solvers)	Plant layout with initial design values Ranges for different design and control parameters Plant model Influent characteristics	Design and control engineers	Moles *et al.* (2003)

(*Continued*)

Table 4.3 Review of uncertainty propagation methods and analysis works applied for model-based evaluation (*Continued*).

DESIGN OPTIMISATION							
Purpose	**Method**	**Accuracy**	**Difficulty**	**Time**	**Data**	**Engineering Task**	**Reference**
Screen WRRF technologies (emerging + state of the art) with a decision-making framework	Stochastic dynamic programming Latin hypercube sampling Orthogonal arrays	Sensitive to performance of emerging technology (i.e., data) Influent characteristics	Decision-making framework clear yet numerical solution complicated	12 167 model evaluations (CPU time not quantified but expected to be long)	Influent data Emerging WRRF technologies and performance data Uncertainty in data	Technology screening purposes hence for design engineers and decision makers	Tsai *et al.* (2004)
Evaluate WRRF system design/upgrade options	MC (method of Rousseau *et al.*, 2001)	Relative measure for comparison1	Medium	Long (weeks), depends on the computation power and the scenario (9 × 100 MC runs)	Models for WRRF configurations Yearly influent profile/load Influent fractions Climate cost index	Researchers and designers (highly relevant) Dedicated software to generate MC samples + run them	Benedetti *et al.* (2006)
Challenge the traditional design approaches in view of future uncertainty of WRRFs: scenario planning for accounting	Historical plant data analysis		Simple	Pending data collection issues	Historical data on influent load, performance, modifications, changes, … Socio-economic development data	Designers	Dominguez and Gujer (2006)
Control alternatives evaluation for WRRF operation	Monte Carlo + multi-criteria decision-making framework	Relative	Medium	Long (weeks)	Models for WRRF configurations Influent profile/load Influent fractions Cost index	Designers and operators, control engineer; for decision making	Flores-Alsina *et al.* (2009); Flores-Alsina *et al.* (2008)

Note: Monte Carlo (MC).

4.5.2 Discussion

4.5.2.1 Model calibration

With respect to how uncertainty affects model calibration, the publications that were reviewed focused on determining wastewater composition and kinetic variables based on available plant data. They present methods relevant to determining the parameters that result in the best fit of the models along with whether the model results are statistically significant.

4.5.2.2 Sensitivity analysis

Several authors focused on the sensitivity of models to the uncertainty of inputs and how that affects the design and performance of treatment facilities. In the simplest approach to sensitivity analysis, the change in model result for a selected output was measured by individually varying input parameters by 10%. Most of these sensitivity analyses were completed using the Monte Carlo approach to varying input parameters. This approach is generally considered more rigorous because it may show the interaction between multiple parameters. As would be expected based on the ASM-based models, each output had a unique set of input parameters that it was most sensitive to. Even when model inputs fell within the expected range, ASMs showed that the uncertainty in the output was significant enough that the ability to meet discharge limits could be uncertain. Similarly, the 2-D clarifier model showed significant uncertainty in secondary clarifier performance due to uncertainty in model inputs.

4.5.2.3 Design optimisation

The studies that focused on robust designs, all generally defined their goal as a design that had the lowest cost, while reliably capable of meeting discharge requirements. The results of these studies generally try to illustrate how increasing or decreasing the amount of money spent changes the risk of being capable of meeting permit requirements.

From the reported studies, Monte Carlo emerges as the most commonly used method of uncertainty analysis when evaluating different WRRF plant design and controller alternatives. While there is no explicit mentioning of how the procedure is applied in these studies, one can infer the following requirements for the uncertainty analysis: (i) a mathematical model describing the system, (ii) uncertainty range and distribution of the parameters in the system (that could be influent data or biochemical parameters). Mostly a uniform distribution is assumed with the upper and lower ranges adopted from literature. There is no standard on the upper and lower range of ASM parameters, (iii) uncertainty analysis typically represented by a cumulative distribution function (CDF) or by a mean accompanied by a standard deviation.

Besides the Monte Carlo method, the following methods are alternatively used (i) generalised likelihood uncertainty estimation (GLUE) which is a Bayesian approach, (ii) fuzzy linear regression method, (iii) stochastic dynamic programming. All these alternative methods add complexity since the user is expected to have some skills and expertise in statistical and numerical programming. It is the authors' opinion that Monte Carlo simulation is intuitively simple hence can be understood by a larger number of practitioners.

4.5.2.4 Computational demand

About the computational demand of uncertainty analysis methods, it depends on the method used. The number of Monte Carlo simulations ranges from 100 to 1000 model evaluations. On the other hand, the GLUE method requires a number of simulations on the order of 10 000. It should be remarked that

Monte Carlo simulations are used just for the purpose of propagating input uncertainty (assumed from expert knowledge) to output uncertainty, however, GLUE method aims to first identify the posterior distribution of parameters (a step which requires many model evaluations in the order of 10 000's) and then propagate this to output uncertainty (this step will be comparable to running a Monte Carlo simulations). Similarly, the stochastic dynamic programming required also on the order of 10 000 simulations. The computationally simplest method appears to be fuzzy linear regression as it involves formulation of linear programming problem with fuzzy inequality constraints for which many effective LP solvers are available.

4.5.2.5 Method accuracy

About the accuracy of the methods, this is difficult to comment on but it is clear that the outcome of an uncertainty analysis depends on how the scenario for the uncertainty analysis is defined, on the framing of the analysis (Sin *et al.*, 2009). This sets the objective (what is the question to be answered) and the boundaries for the analysis, that is which system parameters are included as uncertain, what are the upper and lower ranges selected for each uncertain parameter. In other words, the framing of the analysis reminds the analyst to ask the right question and to set-up the right framework to do so. If the scope of the analysis is set too narrow, the outcome will also be narrow, hence missing out the important implications on the design decisions (the outcome being the right question asked, but the answer is biased). On the other hand, if one sets the scope of the uncertainty analysis too large, then the outcome is likely to be too complex to make sense (as there are too many sources contributing to the decision variable), hence uninformative. While there are still needs for better ways to frame uncertainty analysis, there are already available useful examples on how to setup an uncertainty analysis (see Benedetti *et al.*, 2012; Sin *et al.*, 2009, 2011).

In terms of the purpose of uncertainty analysis, most studies reported in this category aimed at providing decision support for comparing different design alternatives, operation (control) alternatives or technology selection alternatives.

4.6 SUMMARY

4.6.1 Input and parameter uncertainty assessment

- Random errors are characterised using statistical measures of precision such as standard deviation and variance. A common technique for detecting systematic errors in an analytical method or an instrument is to analyse a sample whose composition is accurately known. Another option is to use two independent methods to analyse the same sample as shown by Rieger *et al.* (2005).
- Alternative techniques for detecting and removing systematic errors include multivariate outlier detection methods, statistical process control methods, and data reconciliation. Statistical process control methods and data reconciliation are well suited to on-line applications as they can be easily automated, can simultaneously consider a number of variables, and do not require comparison to a reference method or sample, which may not always be practical. In addition, they can account for auto- and cross-correlation.
- The uncertainty in model parameters is typically assessed as part of parameter estimation. Parameter inference or confidence regions can be developed after parameter estimation based on an approximate variance–covariance matrix for the parameters. This often provides a sufficient approximation of the uncertainty, which can be better approximated using other, more sophisticated techniques.
- Other potentially more powerful techniques include profiling (Bates & Watts, 1988), Monte Carlo analysis, and the use of Bayesian statistics to create a parameter distribution based on prior and

current knowledge (Cox, 2004). Combining Monte Carlo simulation with a parameter estimation algorithm is recommended when a more detailed evaluation of parameter uncertainty is required as it is a powerful and reasonably easy to understand method. It has the disadvantage of requiring a considerable number of simulations.

4.6.2 Model structure uncertainty assessment

- It is largely recognised that wastewater treatment models have structural uncertainty, but methods for quantifying this are generally not available or addressed. This may be an area where considerable additional work is required.
- The level of sophistication and the nature of the state variables included in wastewater treatment models require that those that use them understand and address the limitations of the mathematical approaches used in the models. The completely mixed stirred tank reactor uniform concentration assumption is widely employed in wastewater treatment models but its limitations, which may be more pronounced with the nature of state variables representing storage products and with non-first-order rate expressions, are not generally addressed. In this case, however, the work within the chemical engineering field provides tools to consider the structural uncertainty of this model assumption and the limiting cases of maximum mixing and complete segregation may need to be addressed more often by wastewater treatment modellers.
- The wastewater treatment modelling profession should not become complacent with the Monte Carlo approach to quantify uncertainty. The tools and knowledge of fundamental and applied mathematics should be considered. There is some indication in the literature that different methods for model uncertainty quantification generate different uncertainty results and, therefore, additional understanding and work are required to determine what meaningful uncertainty results are and how they are truly achieved. Research should focus on finding the best methodologies for specific cases and types of analyses.
- There is some indication in the literature that inherent random variability, and hence uncertainty, in wastewater treatment processes may arise from the nature of the systems. The complex, non-linear nature of the systems and the numerous potential competitive and cooperative populations in systems modelled by the wastewater professional may result in dynamical, so-called chaotic, behaviour. The tools of this discipline of mathematics may play a useful role in describing some wastewater treatment systems.

4.6.3 Propagation of uncertainty in model-based decision making

- Although detailed modelling is useful for process engineering, the uncertainty in the inputs and the complexity of the models still result in significant uncertainty in the model outputs.
- Monte Carlo type techniques can be used for design, sensitivity analyses, and calibration. How the Monte Carlo techniques are applied and how the results are interpreted has been approached differently within all of the publications reviewed with no consensus on the best approach.
- Some more mathematically advanced techniques have been applied to process engineering. Although these approaches may improve results, it is unlikely that they can easily be adopted by practicing engineers due to their complexity.
- All approaches for 'robust designs' showed that the uncertainty in some inputs, both wastewater characterisation as well as kinetic parameters, can significantly affect the model predicted rate of

failure. Therefore, the critical inputs need to be identified, and the variability or uncertainty in their values should be defined and used in the modelling exercise.

This section has reviewed methods that have been used to assess and propagate uncertainty in wastewater treatment analyses. Together, these reviews converge to several overarching conclusions:

(1) There is uncertainty in wastewater treatment process model structure, parameters and inputs. Currently, the profession does not have a comprehensive understanding of the extent, impact or relative importance of these contributions.
(2) The Monte Carlo method is the engineering standard method for uncertainty analysis. It is understood widely and works well. However, the community may benefit from the development of further advanced methods and tools.
(3) Framing of uncertainty analysis by asking the right question and setting up the right framework (boundaries of the analysis) are key to arriving at a meaningful and useful result.
(4) Some successful industrial applications were found, but more case studies are needed to realise the benefits of uncertainty analysis methods.

REFERENCES

Abusam A. and Keesman K. (2002). Sensitivity analysis of the secondary settling tank double-exponential function model. *European Water Management Journal*. Official Publication of the European Water Association (EWA), **2002** (7), 1–9.

Afonso P. M. and da Conceição Cunha M. (2007). Robust optimal design of activated sludge bioreactors. *Journal of Environmental Engineering*, **133**(1), 44–52.

Bard Y. (1974). Nonlinear Parameter Estimation. Academic Press, New York and London. ISBN-13: 978-0120782505.

Barker P. and Dold P. (1995). COD and nitrogen mass balances in activated sludge systems. *Water Research*, **29**(2), 633–643.

Bates D. and Watts D. (1988). Nonlinear Regression Analysis and Its Applications. Wiley New York, New York, NY, USA.

Benedetti L., Bixio D. and Vanrolleghem P. A. (2006). Benchmarking of WWTP design by assessing costs, effluent quality and process variability. *Water Science and Technology*, **54**(10), 95–102.

Benedetti L., Batstone D. J., De Baets B., Nopens I. and Vanrolleghem P. A. (2012). Uncertainty analysis of WWTP control strategies made feasible. *Water Quality Research Journal of Canada*, **47**(1), 14–29.

Berthouex P. (1989) Constructing control charts for wastewater treatment plant operation. *Research Journal of the Water Pollution Control Federation*, **61**(9), 1534–1551.

Bertrand-Krajewski J.-L., Winkler S., Saracevic E., Torres A. and Schaar H. (2007). Comparison of and uncertainties in raw sewage COD measurements by laboratory techniques and field UV-visible spectrometry. *Water Science and Technology*, **56**(11), 17–25.

Bixio D., Parmentier G., Rousseau D., Verdonck F., Meirlaen J., Vanrolleghem P. A. and Thoeye C. (2002). A quantitative risk analysis tool for design/simulation of wastewater treatment plants. *Water Science and Technology*, **46**(4–5), 301–307.

Box G. and Draper N. (1965). The Bayesian estimation of common parameters from several responses. *Biometrika*, **52** (3–4), 355.

Box G. and Lucas H. (1959). Design of experiments in non-linear situations. *Biometrika*, **46**(1–2), 77.

Box G., Hunter J. and Hunter W. (1978). Statistics for Experimenters: an Introduction to Design, Data Analysis, and Model Building. John Wiley and Sons Inc., Hoboken, NJ, USA. ISBN-13: 978-0471093152.

Brouwer H., Klapwijk A. and Keesman K. J. (1998). Identification of activated sludge and wastewater characteristics using respirometric batch-experiments. *Water Research*, **32**(4), 1240–1254.

Cameron D. B., Paterson W. R. and Morton W. (1992). Dynamic Data Reconciliation in Process Simulation. Technical Association of the Pulp and Paper Industry (TAPPI) Engineering Conference, Boston, MA, USA.

Checchi N. and Marsili-Libelli S. (2005). Reliability of parameter estimation in respirometric models. *Water Research*, **39**(15), 3686–3696.

Chen H. W. and Chang N. B. (2002). A comparative analysis of methods to represent uncertainty in estimating the cost of constructing wastewater treatment plants. *Journal of Environmental Management*, **65**(4), 383–409.

Chen J., Bandoni A. and Romagnoli J. A. (1997). Robust estimation of measurement error variance/covariance from process sampling data. *Computers & Chemical Engineering*, **21**(6), 593–600.

Cox C. D. (2004). Statistical distributions of uncertainty and variability in activated sludge model parameters. *Water Environment Research*, **76**(7), 2672–2685.

Crowe C. (1996). Data reconciliation – progress and challenges. *Journal of Process Control*, **6**(2–3), 89–98.

Curlin M., Matic M., Matosic M., Mijatovic I. and Kurtanjek Z. (2004). Effects of hydraulic residence time and mixing on wastewater treatment in a membrane bioreactor. *Chemical and Biochemical Engineering Quarterly*, **18**(2), 97–104.

Danckwerts P. (1958). The effect of incomplete mixing on homogeneous reactions. *Chemical Engineering Science*, **8**(1–2), 93–102.

De Pauw D. (2005). Optimal Experimental Design for Calibration of Bioprocess Models: A Validated Software Toolbox. BIOMATH, University of Ghent. Belgium, http://hdl.handle.net/1854/LU-472062 (accessed 2021).

De Pauw D. and Vanrolleghem P. A. (2003). Practical aspects of sensitivity analysis for dynamic models. Proceedings of the 4th IMACS Symposium on Mathematical Modelling (MATHMOD), Vienna, Austria.

Doby T. A., Loughlin D. H., de los Reyes III F. L. and Ducoste J. J. (2002). Optimisation of activated sludge designs using genetic algorithms. *Water Science and Technology*, **45**(6), 187–198.

Dochain D. and Vanrolleghem P. A. (2001). Dynamical Modelling and Estimation in Wastewater Treatment Processes. IWA Publishing, UK, pp. 360. ISBN13: 9781900222501.

Dominguez D. and Gujer W. (2006). Evolution of a wastewater treatment plant challenges traditional design concepts. *Water Research*, **40**(7), 1389–1396.

Draper N. and Smith H. (1989). Applied Regression Analysis, 3rd edn. Wiley, New York, NY, USA.

Duchesne S., Beck M. B. and Reda A. (2001). Ranking stormwater control strategies under uncertainty: the river Cam case study. *Water Science and Technology*, **43**(7), 311–320.

Flores-Alsina X., Rodriguez-Roda I., Sin G and Gernaey K. V. (2008). Multi-criteria evaluation of wastewater treatment plant control strategies under uncertainty. *Water Research*, **42**(17), 4485–4497.

Flores-Alsina X., Rodriguez-Roda I., Sin G and Gernaey K. V. (2009). Uncertainty and sensitivity analysis of control strategies using the benchmark simulation model No1 (BSM1). *Water Science Technology*, **59**(3), 491–499.

Franks S. W. (1999). Disaggregation of environmental factors affecting sewer pipe failures. *Journal of Infrastructure Systems*, **5**(4), 150–158.

Friedler E. and Butler D. (1996). Quantifying the inherent uncertainty in the quantity and quality of domestic wastewater. *Water Science and Technology*, **33**(2), 65–78.

Graham D., Knapp C., Van Vleck E. S., Bloor K., Lane T. B. and Graham C. E. (2007). Experimental demonstration of chaotic instability in biological nitrification. *The ISME Journal*, **1**(5), 385–393.

Griborio A., McCorquodale J., Pitt P., Vinci P. and Wang T. (2007). Uncertainty analysis in wastewater engineering modelling: understanding the limitations. *Proceedings of the Water Environment Federation*, **2007**(17), 1584–1596.

Guay M. and McLean D. (1995). Optimization and sensitivity analysis for multiresponse parameter estimation in systems of ordinary differential equations. *Computers and Chemical Engineering*, **19**(12), 1271–1286.

Gujer W. (2002). Microscopic versus macroscopic biomass models in activated sludge systems. *Water Science and Technology*, **45**(6), 1–11.

Haider S., Svardal K., Vanrolleghem P. A. and Kroiss H. (2003). The effect of low sludge age on wastewater fractionation (SS, SI). *Water Science and Technology*, **47**(11), 203–209.

Huo J., Robinson R. B. and Cox C. D. (2004). Uncertainty and Sensitivity Analysis of Activated Sludge Model No. 1 by Monte Carlo Simulation for Single CSTR with Universal Distribution Parameters. EWRI'04 World Water & Environmental Resources Congress, June 27–July 1, 2004, Salt Lake City, Utah, USA.

Huo J., Jiang Y., Seaver W., Robinson R. B. and Cox C. D. (2006). Statistically based design of wastewater treatment plants (WWTPs) using Monte Carlo simulation of Activated Sludge Model No.1 (ASM1). World Environmental and Water Resources Congress, Omaha, Nebraska, USA, ASCE.

Ibrahim G., Habib H. and Saleh O. (2008). Periodic and chaotic solutions for a model of a bioreactor with cell recycle. *Biochemical Engineering Journal*, **38**(2), 124–137.

Joannis C., Ruban G., Gromaire M. C., Cheboo G. and Bertrand-Krajewski J. L. (2008). Reproducibility and uncertainty of wastewater turbidity measurements. *Water Science and Technology*, **57**(10), 1667–1673.

Kletting M., Rauh A., Aschemann H. and Hofer E. P. (2007). Consistency tests in guaranteed simulation of nonlinear uncertain systems with application to an activated sludge process. *Journal of Computational and Applied Mathematics*, **199**(2), 213–219.

Koch G., Kuhni M. and Siergrist H. (2001). Calibration and validation of an ASM3-based steady-state model for activated sludge systems-part I: prediction of nitrogen removal and sludge production. *Water Research*, **35**(9), 2235–2245.

Kops S. and Vanrolleghem P. A. (1996). An evaluation of methodologies for uncertainty analysis in biological waste water treatment. 9th European Junior Scientist Workshop, 11–14 April, Klive, UK.

Kourti T. and MacGregor J. (1995). Process analysis, monitoring and diagnosis, using multivariate projection methods. *Chemometrics and Intelligent Laboratory Systems*, **28**(1), 3–21.

Krasnochtanova I. A. R., Kletting M., Aschemann H., Hofer E. P. and Schoop K. M. (2009). Interval methods as a simulation tool for the dynamics of biological wastewater treatment processes with parameter uncertainties. *Applied Mathematical Modelling*, **34**(3), 744–762.

Lavallee B., Lessard P. and Vanrolleghem P. A. (2005). Modelling the metabolic adaptations of the biomass under rapid growth and starvation conditions in the activated sludge process. *Journal of Environmental Engineering and Science*, **4**(6), 533–548.

Lee T. T., Wang F. Y. and Newell R. B. (1999a). Distributed parameter approach to the dynamics of complex biological processes. *AIChE Journal*, **45**(10), 2245–2261.

Lee T. T., Wang F. Y. and Newell R. B. (1999b). On the modelling and simulation of a BNR activated sludge process based on distributed parameter approach. *Water Science and Technology*, **39**(6), 79–88.

Leis J. and Kramer M. (1988). The simultaneous solution and sensitivity analysis of systems described by ordinary differential equations. *ACM Transactions on Mathematical Software (TOMS)*, **14**(1), 45–60.

Lennox J. and Rosen C. (2002). Adaptive multiscale principal components analysis for online monitoring of wastewater treatment. *Water Science & Technology*, **45**(4), 227–235.

Levenspiel O. (1999). Industrial and Chemistry Engineering Research. ACS Publications, DC, USA.

Makinia J. and Wells S. (2000a). A general model of the activated sludge reactor with dispersive flow – I. Model development and parameter estimation. *Water Research*, **34**(16), 3987–3996.

Makinia J. and Wells S. A. (2000b). A general model of the activated sludge reactor with dispersive flow – II. Model verification and application. *Water Research*, **34**(16), 3997–4006.

Marsili-Libelli S. and Tabani F. (2002). Accuracy analysis of a respirometer for activated sludge dynamic modelling. *Water Research*, **36**(5), 1181–1192.

Meijer S., Van Der Spoel H., Susanti S. and Heijne J. J. (2002). Error diagnostics and data reconciliation for activated sludge modelling using mass balances. *Water Science and Technology*, **45**(6), 145–156.

Moles C. G., Gutierrez G., Alonso A. A. and Banga J. R. (2003). Integrated process design and control via global optimisation: a wastewater treatment plant case study. *Chemical Engineering Research and Design*, **81**(5), 507–517.

Montgomery D. and Mastrangelo C. (1991). Some statistical process control methods for autocorrelated data. *Journal of Quality Technology*, **23**(3), 179–193.

Neumann M. B. and Gujer W. (2008). Underestimation of uncertainty in statistical regression of environmental models: influence of model structure uncertainty. *Environmental Science and Technology*, **42**(11), 4037–4043.

Nowak O., Franz A., Svardal K., Müller V. and Kühn V. (1999). Parameter estimation for activated sludge models with the help of mass balances. *Water Science and Technology*, **39**(4), 113–120.

Omlin M. and Reichert P. (1999). A comparison of techniques for the estimation of model prediction uncertainty. *Ecological Modelling*, **115**(1), 45–59.

Petersen B. (2000). Calibration, identifiability and optimal experimental design of activated sludge models. PhD thesis, Department of Applied Mathematics, Biometrics and Process Control (BIOMATH) University of Ghent, Belgium.

Petersen B., Gernaey K. and Vanrolleghem P. A. (2000). Improved theoretical identifiability of model parameters by combined respirometric–titrimetric measurements. A generalisation of results. Proceedings IMACS 3rd Symposium on Mathematical Modelling, Vienna, Austria.

Portielje R., Hvitved-Jacobsen T. and Schaarup-Jensen K. (2000). Risk analysis using stochastic reliability methods applied to two cases of deterministic water quality models. *Water Research*, **34**(1), 153–170.

Puig S., van Loosdrecht M., Colprim J. and Meijer S. C. F. (2008). Data evaluation of full-scale wastewater treatment plants by mass balance. *Water Research*, **42**(18), 4645–4655.

Rauh A., Kletting M., Aschemann H. and Hofer E. P. (2004). Application of interval arithmetic simulation techniques to wastewater treatment processes. Proceedings of the IASTED International Conference on Modelling, Identification, and Control (MIC 2004), Grindelwald, Switzerland, pp. 287–293. https://www.actapress.com/Content_of_Proceeding.aspx?proceedingID=247 (accessed 2021).

Rauh A., Kletting M., Aschemann H. and Hofer E. P. (2007). Reduction of overestimation in interval arithmetic simulation of biological wastewater treatment processes. *Journal of Computational and Applied Mathematics*, **199**(2), 207–212.

Rawlings J. and Ekerdt J. (2002). Chemical Reactor Analysis and Design Fundamentals. Publisher: Nob Hill Publishing, Madison, Wisconsin, USA.

Rieger L., Thomann M., Gujer W. and Siegrist H. (2005). Quantifying the uncertainty of on-line sensors at WWTPs during field operation. *Water Research*, **39**(20), 5162–5174.

Rieger L., Takacs I., Villez K., Siegrist H., Lessard P., Vanrolleghem P. A. and Comeau Y. (2010). Data reconciliation for wastewater treatment plant simulation studies planning for high-quality data and typical sources of errors. *Water Environment Research*, **82**(5), 426–433.

Rieger L., Gillot S., Langergraber G., Ohtsuki T., Shaw A., Takacs I. and Winkler S. (2013). Guidelines for Using Activated Sludge Models, Scientific and Technical Report No. 22, International Water Association (IWA), London, UK.

Robinson R. B., Cox C. D. and Odom K. (2005). Identifying outliers in correlated water quality data. *Journal of Environmental Engineering*, **131**(4), 651–657.

Romagnoli J. and Sanchez M. (2000). Data Processing and Reconciliation for Chemical Process Operations. Academic Press, San Diego, CA, USA.

Rousseau D., Verdonck F., Moerman O., Carrrette R., Thoeye C., Meirlaen J. and Vanrolleghem P. A. (2001). Development of a risk assessment based technique for design/retrofitting of WWTPs. *Water Science and Technology*, **43**(7), 287–294.

Saikaly P. and Oerther D. (2004). Bacterial competition in activated sludge: theoretical analysis of varying solids retention times on diversity. *Microbial Ecology*, **48**(2), 274–284.

Schuler A. (2005). Diversity matters: dynamic simulation of distributed bacterial states in suspended growth biological wastewater treatment systems. *Biotechnology and Bioengineering*, **91**(1), 62–74.

Schuler A. (2006). Process hydraulics, distributed bacterial states, and biological phosphorus removal from wastewater. *Biotechnology and Bioengineering*, **94**(5), 909–920.

Sin G. (2004). Systematic calibration of activated sludge models. PhD thesis, Department of Applied Mathematics, Biometrics and Process Control (BIOMATH), University of Ghent, Belgium.

Sin G. and Vanrolleghem P. A. (2006). Evolution of an ASM2s-like model structure due to operational changes of an SBR process. *Water Science and Technology*, **53**(12), 237–245.

Sin G., De Pauw D. J. W., Weijers S. and Vanrolleghem P. A. (2008). An efficient approach to automate the manual trial and error calibration of activated sludge models. *Biotechnology and Bioengineering*, **100**(3), 516–528.

Sin G., Gernaey K. V., Neumann M. B., van Loosdrecht M. C. M. and Gujer W. (2009). Uncertainty analysis in WWTP model applications: a critical discussion using an example from design. *Water Research*, **43**(11), 2894–2906.

Sin G., Gernaey K. V., Neumann M. B., van Loosdrecht M. C. M. and Gujer W. (2011). Global sensitivity analysis in wastewater treatment plant model applications: prioritising sources of uncertainty. *Water Research*, **45**(2), 639–651.

Skoog D. A., West D. M. and Holler F. J. (1995). Fundamentals of Analytical Chemistry, 7th edn. Saunders College Publishing, USA. ISBN-13: 978-0-03-005938-4.

Standard Methods for the Examination of Water and Wastewater. (1998). 20th edn. American Public Health Association/American Water Works Association/Water Environment Federation, Washington DC, USA.

Stroot P., Saikaly P. and Oerther D. B. (2005). Dynamic growth rates of microbial populations in activated sludge systems. *Journal of Environmental Engineering*, **131**, 1698.

Tansel B. (1999). Propagation of parameter uncertainties to system dependability. *Civil Engineering and Environmental Systems*, **16**(1), 19–35.

Tenno R. and Uronen P. (1995). State and parameter estimation for wastewater treatment processes using a stochastic model. *Control Engineering Practice*, **3**(6), 793–804.

Thomann M. (2008). Quality evaluation methods for wastewater treatment plant data. *Water Science and Technology*, **57**(10), 1601–1609.

Thomann M., Rieger L., Frommhold S. and Siegrist H. (2002). An efficient monitoring concept with control charts for on-line sensors. *Water Science and Technology*, **46**(4–5), 107–116.

Tomita R., Park S. and Sotomayor O. (2002). Analysis of activated sludge process using multivariate statistical tools – a PCA approach. *Chemical Engineering Journal*, **90**(3), 283–290.

Tsai J. C. C., Chen V. C. P., Beck M. B. and Chen J. (2004). Stochastic dynamic programming formulation for a wastewater treatment decision-making framework. *Annals of Operations Research*, **132**(1–4), 207–221.

Van der Heijden R., Heijnen J., Hellinga C., Romein B. and Luyben K. C. (1994a). Linear constraint relations in biochemical reaction systems: I. Classification of the calculability and the balanceability of conversion rates. *Biotechnology and Bioengineering*, **43**(1), 3–10.

Van der Heijden R., Heijnen J., Hellinga C., Romein B. and Luyben K. C. (1994b). Linear constraint relations in biochemical reaction systems: II. Diagnosis and estimation of gross errors. *Biotechnology and Bioengineering*, **43**(1), 11–20.

Van der Heijden R., Heijnen J., Hellinga C., Romein B. and Luyben K. C. (1994c). Linear constrain relations in biochemical reaction systems III. Sequential application of data reconciliation for sensitive detection of systematic errors. *Biotechnology and Bioengineering*, **44**(7), 781–791.

Vanrolleghem P. A. and Keesman K. J. (1996). Identification of biodegradation models under model and data uncertainty. Proceedings of the 1995 IAWQ Interdisciplinary Symposium on Uncertainty, Risk and Transient Pollution Events, 26–28 July 1995, Pergamon Press Inc., Exeter, UK.

Vanrolleghem P. A., Van Daele M. and Dochain D. (1995). Practical identifiability of a biokinetic model of activated sludge respiration. *Water Research*, **29**(11), 2561–2570.

Vanrolleghem P. A., Insel G., Petersen B., Sin G., Nopens I., Dovermann H., Weijers S. and Gernaey K. (2003). A comprehensive model calibration procedure for activated sludge models. Proceedings of the Water Environment Federation, Alexandria, VA, USA. Water Environment Federation 76th Annual Technical Exhibition and Conference (WEFTEC) 2003, Los Angeles, CA, USA.

Von Sperling M. (1993). Parameter estimation and sensitivity analysis of an activated sludge model using Monte Carlo simulation and the analyst's involvement. Proceedings of the 6th JAWO Workshop on Instrumentation, Control and Automation of Water and Wastewater Treatment and Transport Systems, 17–25 June 1993, Banff, Canada, Published by Pergamon Press Inc.

Zhang Y. and Henson M. (2001). Bifurcation analysis of continuous biochemical reactor models. *Biotechnology Progress*, **17**(4), 647–660.

Zwietering T. (1959). The degree of mixing in continuous flow systems. *Chemical Engineering Science*, **11**(1), 1–15.

Chapter 5

The DOUT uncertainty analysis methodology – combining models, statistics and design guidelines

5.1 INTRODUCTION

To facilitate the transition of wastewater treatment plant design from guidelines (heuristic) and deterministic approaches to a probabilistic approach, models need to be combined with statistical methods. Models consider the plant as a system and take into account the interdependencies within the process train. Statistical methods can be used to propagate for variability and uncertainty to the model outputs.

Success in transitioning to a probabilistic design requires the development of a set of protocols to guide the design engineer through the process with the maximum degree of objectivity and transparency. This will enable stakeholders with the proper expertise to understand the rationale for the design and how the uncertainties in the design were handled. In each engineering project phase (planning, preliminary design, detailed design, operation), modelling can be applied to support the design decisions and evaluate specific uncertainties.

This chapter discusses the steps and key elements of such a methodology, based on the work of the Task Group. The work presented in this chapter is largely based on Talebizadeh (2015).

5.2 THE INCLUSION OF UNCERTAINTY ANALYSIS IN A MODEL-BASED PROJECT

5.2.1 General tasks

There are several published uncertainty analysis protocols, most of them developed by researchers working in the water resources field (e.g., Jakeman *et al.*, 2006; Refsgaard *et al.*, 2007).

Figure 5.1 shows a list of tasks that need to be incorporated in a model-based project when uncertainty analysis is undertaken.

 Uncertainty in Wastewater Treatment Design and Operation: Addressing Current Practices and Future Directions
Editor(s): Evangelia Belia, Lorenzo Benedetti, Bruce Johnson, Sudhir Murthy, Marc Neumann, Peter Vanrolleghem and Stefan Weijers
doi: 10.2166/9781780401034_0071

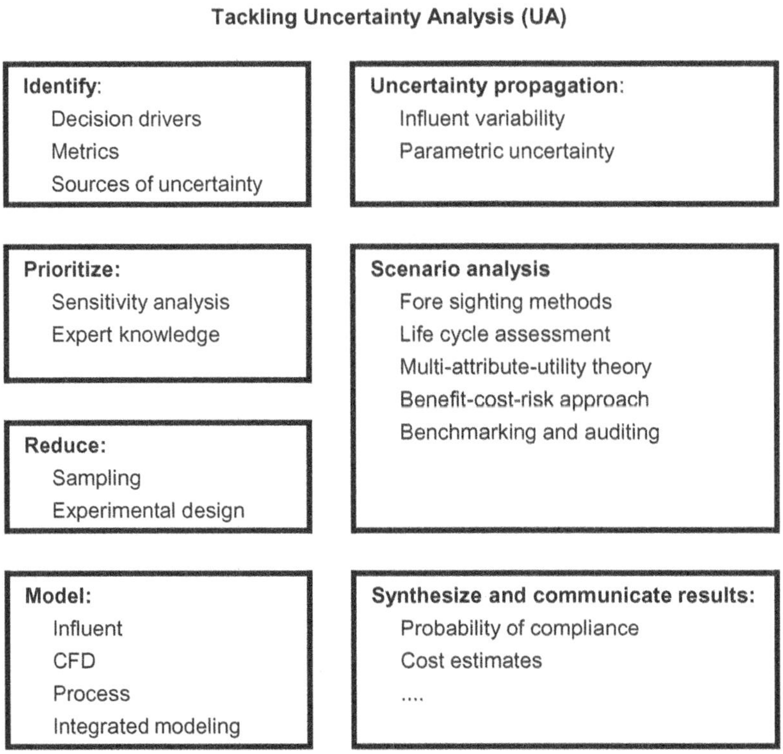

Figure 5.1 List of tasks in a model-based project with uncertainty analysis (adapted from Jakeman *et al.*, 2006).

5.2.2 Linking process modelling steps and uncertainty methodology tasks

Figure 5.2 identifies which uncertainty-related tasks need to be considered at each stage of a wastewater treatment modelling project if uncertainty and variability are to be evaluated. The typical six-step simulation-based project execution flow sheet (Rieger *et al.*, 2013) has been used as a basis for a modelling project.

At the ***project definition*** step, the various sources of uncertainty need to be identified. Depending on the boundaries of the project some of these uncertainties may be associated with external non-controllable sources. It is crucial that sources of uncertainty are considered explicitly early in the modelling study. However, at this stage uncertainties are seldom quantified. It is also at this stage that the expected model performance/predictive accuracy needs to be established. The accuracy performance criteria will be reconsidered during the modelling process to assess whether the original expectations were realistic based on the available data and model capabilities.

During the ***data collection and reconciliation*** step, extra experiments or measurements can be taken to reduce the uncertainty of certain sources. The statistical description of key sources of uncertainty can be performed at this stage.

At the ***model set-up*** phase the modeller typically selects the model category (e.g., ASM1, ASM2d) and configuration of the process train (e.g., number of tanks, number of trains to be modelled). These decisions will determine the model structure uncertainty of the simulated process.

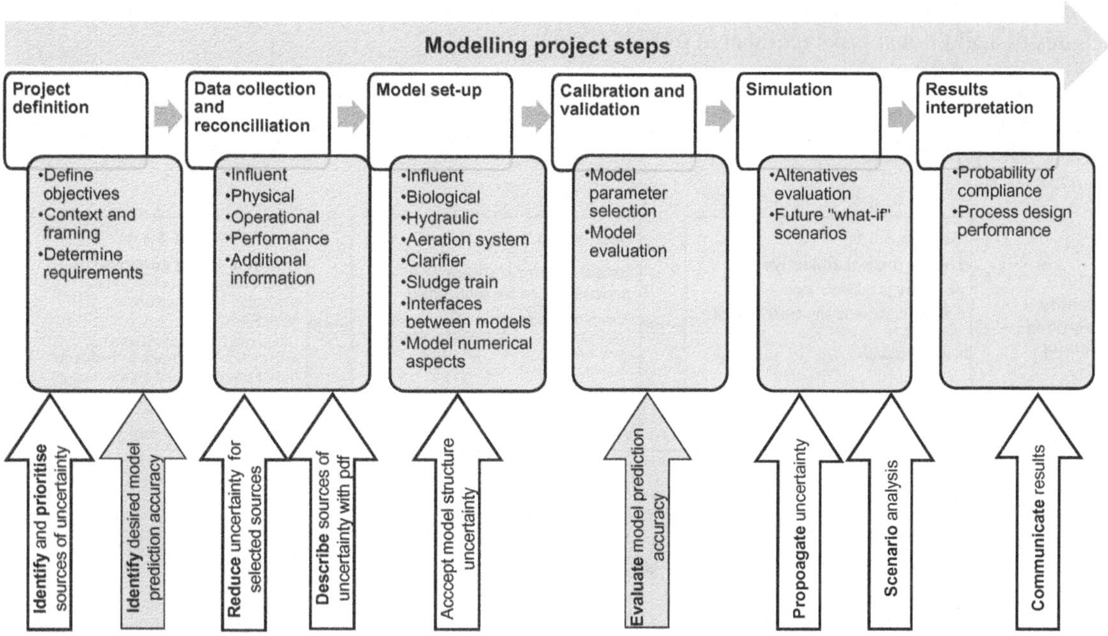

Figure 5.2 Modelling project steps and uncertainty evaluation tasks.

Towards the end of the ***calibration and validation*** step, an assessment is undertaken that compares calibration with validation results. The validation step is performed with an independent data set that contains none of the data used for calibration. This task evaluates possible biases in the model and assesses whether the model performance is good enough to meet the expected accuracy requirements. The calibration and validation steps are only relevant for existing plants going through an upgrade. Models for green-field sites where the plant does not exist, cannot be calibrated/validated. In the ***simulation*** step, uncertainty assessment and propagation are conducted. Uncertainty propagation is often limited to quantitative uncertainty. However, scenario uncertainties can also be taken into account in this step (Refsgaard *et al.*, 2007). For definitions of quantifiable and scenario uncertainty, see Section 1.3.2.

During each step of the modelling project an evaluation is performed to decide whether there are sufficient data to proceed with the modelling, whether the uncertainty in the model is at a level where the results can be expected to be useful, whether the assumptions made in the model are realistic and how the study outcome may be influenced by the implicit and explicit assumptions made in the model (Refsgaard *et al.*, 2007).

5.3 BRIDGING DESIGN GUIDELINES AND STEADY-STATE DESIGN WITH DYNAMIC STOCHASTIC MODELLING

Part of the work of the Task Group was to develop a methodology that incorporates explicit uncertainty evaluations specifically for wastewater plant design. The goal of this methodology is to provide design engineers with a quantitative probability of compliance for the design under consideration. This section describes the proposed design methodology in a typical design project.

Figure 5.3 shows the tasks to be followed in the proposed protocol. The Task Group methodology includes two additional tasks (compared to Figure 5.1).

(1) An evaluation of a set of preliminary designs based on design guidelines (e.g., ATV (2000), Metcalf & Eddy *et al.* (2013)).
(2) An optimisation loop.

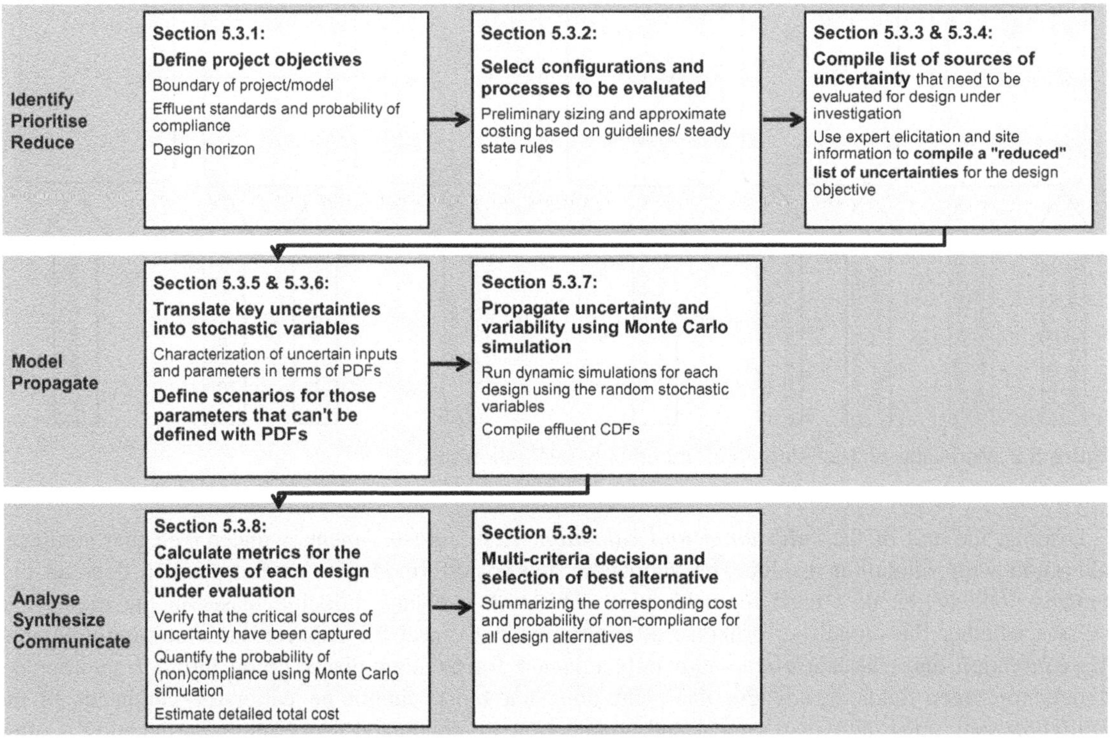

Figure 5.3 Flow sheet of the Task Group methodology for probabilistic plant design. Abbreviations: PDF = probability density function, CDF = cumulative frequency distribution (adapted from Talebizadeh, 2015).

5.3.1 Define project objectives

At the start of a project the specific objectives, design constraints and the boundaries of the system should be clearly defined. Defining the system boundaries will define the inputs to the model, the sources of variability and uncertainty and the type of analysis required to evaluate the performance of each design alternative.

5.3.2 Select configurations to be evaluated

5.3.2.1 Generation of a set of pre-designs with different levels of safety

In this task a set of pre-designs is generated. Each pre-design incorporates project specifications, available data and other prior knowledge. These pre-design alternatives can be generated using design guidelines (see Chapter 2, Section 2.2.1) or by running a steady-state model.

In contrast to current practice where single values are selected for design inputs to generate the size of different treatment units, in this methodology, each design input is described with a uniform distribution. The lower and upper limits of the distributions are based on expert opinion, previous studies or available data. The various pre-designs are then generated by randomly sampling (Monte Carlo) the distributions. As a result, each pre-design has a specific level of conservatism resulting from the random selection of design inputs.

Figure 5.4 shows a schematic representation of this step. The grey areas represent the uniform distribution of each input and of the output (design alternative). The white, dark grey and black lines represent examples of three distinct levels of conservatism.

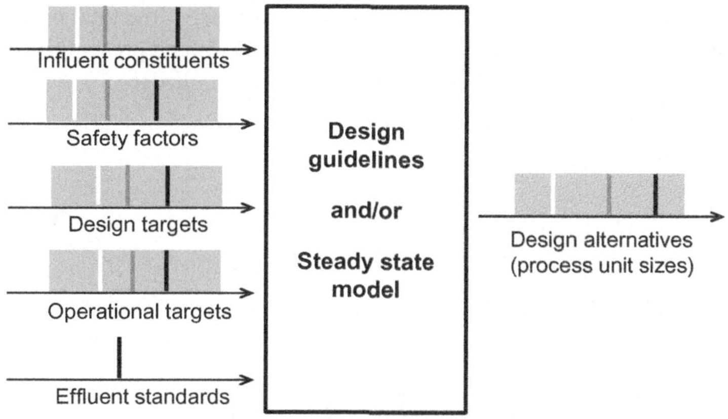

Figure 5.4 Generating a set of design alternatives using different input values sampled from uniform distributions of each of the inputs (Talebizadeh, 2015).

Figure 5.4 also shows the categories of the design inputs typically required to generate a pre-design using design guidelines (or a steady-state model):

- Influent wastewater constituents derived from the basis of design (e.g., flow, COD, ammonia);
- Safety factors applied to design parameters such as biomass growth rates, influent concentrations or effluent standards;
- Unit process design targets (e.g., overflow rates for clarifiers, SVI);
- Operational targets which include utility-specific wishes or constraints relating to the operation of the treatment plant;
- Effluent standards (fixed values with no uncertainty attached to them) which include the effluent concentration values that are to be met.

Correlations between inputs need to be taken into account by including correlation relationships. Alternatively, a correlated sampling technique which produces correlated input variables can be used (Iman & Conover, 1982).

Finally, the number of Monte Carlo simulations to be performed should be large enough to ensure that the entire space of potential design outputs (i.e., the size of different treatment units) is covered.

5.3.2.2 Screening of pre-designs

A number of the design alternatives generated in the previous step may not be feasible due to site-specific constraints and many may not be significantly different from each other in size and performance. In addition to plant performance, a preliminary cost estimate (Gillot *et al.*, 1999) might identify and eliminate pre-designs with prohibitive capital or operational costs.

Non-feasible, very similar or very high-cost design alternatives should be removed and only a small number of pre-designs that are representative of the design space generated in the previous step should be analysed further.

An efficient way to reduce the number of generated pre-designs is through the application of a *K*-means clustering method (Hartigan & Wong, 1979). The *K*-means algorithm aims to partition a dataset into *K* clusters by minimising the sum of distances between all values in a cluster and the respective centroid. The cluster centroids are representative pre-designs to be analysed further.

Figure 5.5 shows an example of the implementation of the *K*-means algorithm for a hypothetical set of pre-designs. The light grey dots plot the area of the secondary clarifier vs. the total bioreactor volume for the several hundred pre-designs generated for this example. The black dots represent the cluster centroids that were determined based on a *K*-means clustering method with seven clusters.

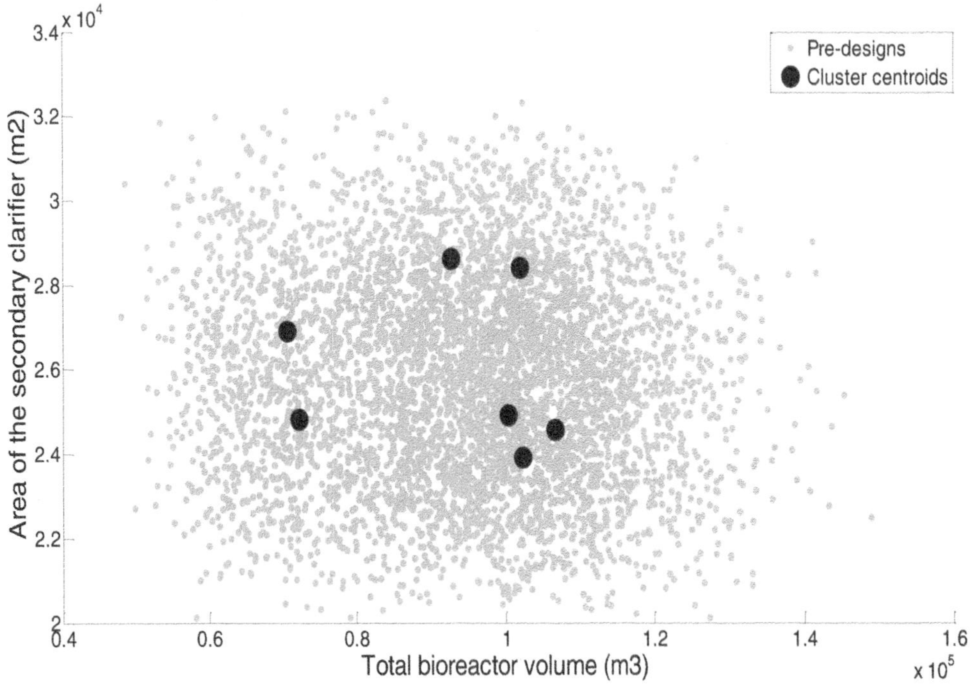

Figure 5.5 Generated pre-designs and cluster locations for a *K*-means clustering with seven clusters (Talebizadeh, 2015).

In addition to the pre-designs identified with the *K*-means clustering method (the centroids), the design engineer may choose additional pre-designs, for various reasons (e.g., to evaluate performance under extreme influent conditions), to be included in the group subjected to further analysis.

5.3.2.3 Preliminary evaluation of pre-designs with dynamic data

The preliminary design alternatives chosen from the analysis discussed in Section 5.3.2.2 are then evaluated under dynamic influent conditions. A one-year long influent time series representative of the expected variability in influent conditions is developed for this purpose. The model is run with this time series as input using the model's set of default parameters. However, certain parameters can be adjusted, if information is available that indicates that the default values do not apply for the facility under design. The output from the simulation is taken as an indicator of the expected performance behaviour of the plant under dynamic conditions.

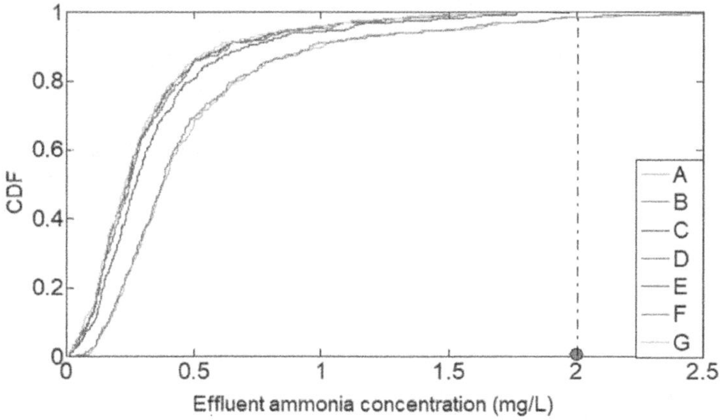

Figure 5.6 Example of effluent CDFs for seven pre-design alternatives. The performance of each of the pre-designs is evaluated against the effluent permit. In this case 2 mg/L of ammonia (Talebizadeh, 2015).

Cumulative distribution functions (CDFs) can be constructed for each preliminary design for the critical effluent water parameters as shown in Figure 5.6. The CDFs are examined to eliminate those design alternatives with a poor performance in terms of effluent quality under dynamic conditions. In addition, the comparison can serve as a tool for eliminating design alternatives having the same treatment performance. This step reduces even further the number of pre-designs that receive further consideration.

5.3.3 Identify sources of variability and uncertainty to be evaluated

In the step described above, several pre-designs are selected for further evaluation. These selected pre-designs are subjected to dynamic probabilistic analysis to generate a profile of their expected performance. In the previous step, the dynamic simulation of the selected alternatives used an influent time series representative of the expected variability in influent conditions and default model parameters. The dynamic probabilistic evaluation in this next step, involves running simulations with stochastic influent and model parameter values (Table 5.1).

To prepare for this, the sources of uncertainty that will be propagated through the dynamic model must be determined. As a first step, a list of the relevant sources of uncertainty is composed. Of all the potential sources of uncertainty only the ones considered by the design engineer critical for the specific objective are evaluated. An uncertainty matrix specific for the project can be constructed along the lines of Tables 5.2–5.4.

The uncertainty matrix lists the location of uncertainty and each source, its nature (variability: irreducible/ epistemic: reducible) and its level (statistical or scenario). Additional discussion on the uncertainty matrix can be found in Walker *et al.* (2003).

Table 5.1 Location (source) of uncertainty in WWTP modelling (after Belia *et al.*, 2009).

Location	Details	Sources	Examples
Inputs	Measured data	Influent data	Current and future predicted flow, COD, ammonia
		Physical data	Tank volume and geometry
		Operational settings	DO set points
		Performance data	Effluent data, reactor concentrations
		Additional info	Input from connected systems for example, sewers, catchment
Model structure and parameters	Model parameters	Hydraulic	Number of tanks in series
		Biokinetic	Maximum growth rates
		Settling	Settling coefficients
	Models	Equations describing processes	Influent model, hydraulic model, aeration system model, process models (biological, settling, etc.)
	Interfaces between models	Mapping of state variables	Waste-activated sludge pumped to an anaerobic digester; digester effluent pumped to sludge treatment
Numerics	Software (model technical aspects)	Solver settings	Wrong solver used (solver does not converge)
		Numerical approximations	Mistakes in model coding
		Software limitations	
		Bugs	
Model output	Propagation of uncertainty	All model uncertainties	Probability of meeting effluent criteria

Chapter 3, Section 3.5, discussed key challenges in transitioning from the guidelines/safety factor design approach to probabilistic design. One of the major challenges identified in the work of the Task Group was that, in the wastewater field, when using models to quantify risk, researchers often focused on a few model parameters. Furthermore, the identification of uncertainties for use in risk analysis tended to be done in an ad-hoc manner. To remedy these weaknesses, and to create a more structured framework for uncertainty analysis the Task Group compiled and categorised the most important sources of uncertainty that impact a wastewater project (Table 5.1). Though this list is not exhaustive, it should cover most of the sources of uncertainty of interest.

5.3.3.1 Input variability and uncertainty

Table 5.2 lists the sources of variability and uncertainty introduced during the data input phase of a modelling project and classifies them as reducible or irreducible. In most cases, uncertainties in input data that are irreducible can be attributed to the inherent variability of the system being modelled, especially influent data. As noted in Table 5.2, uncertainty related to measurement/sampling/reporting errors is reducible. Physical data needed for the model is in most cases known. Uncertainties related to physical information can be irreducible due to unknown factors or reducible. For example, uncertainty regarding the active tank volume of an anaerobic digestion that has sediment accumulation can be reduced through tracer studies. Operating settings and performance data can include some inherent levels

Table 5.2 Examples of input variability and uncertainty.

Location	Details	Nature of Uncertainty	Examples
Influent data	Flow rate, concentrations, influent characterisation data, temperature, data from other models and other systems like sewers	**Irreducible:** due to the inherent variability of the real system like weather, unexpected demographic changes, unexpected factory shutdowns	Increase in influent TKN in new developments due to low-water fixtures and water conservation. New community development with no existing wastewater to characterise or treatment plant to calibrate. Industrial toxic effects on microbial community in treatment plant.
		Reducible: due to data collection for example, sampling method, location, frequency, accuracy of sensors, accuracy of analytical techniques	Plants historically collecting influent BOD, with limited to no COD data. No measurement of influent temperature. Limited influent N and P data available because plant was not required to remove TN or TP, but new permits have TN and TP limits. Biodegradability of industrial waste. C/N/P ratios of industrial waste. Scenario: Future predicted flow, rate of growth, ….
Physical data	Process flow diagram, active (effective) tank volumes, clarifier surface areas, flow splits	**Irreducible:** due to the dynamic behaviour of structures to flow splits and flow changes	Scenario: CFD or tracer studies to determine flow split under different scenarios, however variability in flow splits may remain irreducible.
		Reducible: due to incorrect physical information provided	Unknown true volume constructed or operational depth of structures. Take field measurements if inadequate as-builts provided. Reduction in tank volume due to sediment/grit accumulation. *Influent of plant construction* (e.g., *circular* tanks with *non-ideal* flows and hydrodynamic inefficiencies).
Operational settings	Controller set-points, valve positions, pumped flows	**Reducible:** due to actions different from planned or changes not logged	Change in set-points, incorrect controller set-up (e.g., scales different between field and control room). Uncalibrated controller or instrument (e.g., DO probes, nutrient analysers). Insufficient logging of operator actions (e.g., turning on or off a pump without mentioning the flow rate).
Operational data	Plant performance data	**Reducible:** due to sampling and data collection issues	Composite vs. grab sampling of for example, MLSS. Changes during sample collection and storage can impact performance data. Continuous vs. intermittent pumping. If a pump operates intermittently, the concentration of the TSS on primary sludge can vary if sample taken at beginning or end of pumping cycle. Poorly functioning online equipment.
Equipment performance	Equipment failures	**Irreducible:** for example, due to unexpected equipment failures	Mixers not working in aeration system because limited screening allowed rags into tanks, resulting in different mixing conditions throughout tank.
		Reducible: provide redundant design of critical pieces of equipment and processes	Provide standby unit processes or pieces of equipment to ensure continues operation (e.g., aeration blowers, secondary clarifiers, filers)

Table 5.3 Examples of model structure and model parameter uncertainty.

Location	Details	Nature of Uncertainty	Examples
Influent model	Influent dynamics, characteristics, influent fractions	**Reducible:** due to simplifications of influent dynamics and influent characteristics	Generic diurnal patterns Fixed ratios for influent fractions Limited data used to determine influent fractions
Biological model	**Model structure:** ASM and ADM type models for processes. Type of mathematical expression used to describe processes	**Reducible:** due to simplifications in model structure, choice of mathematical description of processes	ASM models are approximation of reality Biological active species in the wastewater Monod vs. enzymatic kinetics Processes not included or included in simplified form Removal of particulate and colloidal fractions assumed as an instantaneous process One-step vs. two-step nitrification
	Model parameters: growth rates, yields and half saturations are fixed, a priori chosen, calibrated, time varying	**Reducible:** due to lack of knowledge of the appropriate value	Toxic components New process elements that are not properly characterised in literature with poorly defined parameters.
Hydraulic model	**Model structure:** transport and mixing processes, number of trains, number of tanks in series	**Reducible:** due to the simplification of transport and mixing processes in models, inadequate spatial resolution	CSTRs vs. plug flow. Selection of number of trains to model, number of tanks in series.
	Model parameters: empirical hydraulic loading relationships fixed, a priori chosen, calibrated, time varying		Loading on filters are based on simple hydraulic parameters resulting in poor prediction of performance.
Aeration system model	**Model structure:** gas transfer processes, mechanical system details	**Reducible:** due to the simplification of gas transfer processes and aeration system.	Improper model structure for gas–liquid transfer
	Model parameters: fixed, a priori chosen, calibrated, time varying		Parameters such as alpha factors and oxygen transfer efficiency not known

(Continued)

Table 5.3 Examples of model structure and model parameter uncertainty (*Continued*).

Location	Details	Nature of Uncertainty	Examples
Clarifier model	**Model structure:** separation processes, calculation of composite variables and type of mathematical expression used to describe processes.	**Reducible:** due to simplifications in model structure, processes omitted, processes included in simplified form, choice of mathematical description of processes.	Model selection may not sufficiently address solids transport. 1-D, 2-D, CFD analysis
	Model parameters: fixed, a priori chosen, calibrated, time varying	**Irreducible:** due to inherently varying biomass settling properties. **Reducible:** due to our lack of knowledge of the appropriate value.	Selection of design settling characteristics and how they can be related to historical values expressed as the sludge volume index (SVI)
Controllers in plant operations	Control loops, sensors, actuators, time variation of set-points	**Reducible:** due to the oscillation of the aeration system, time delays in control loops, non-linearity of actuators.	Approximate PID tuning values
Interfaces between models	Use of one or several sets of state variables, calculation of composite variables	**Reducible:** due to the aggregation of state variables.	Incompatibility between output of one model and input of another model

Table 5.4 Examples of numerical uncertainty.

Location	Details	Nature of Uncertainty	Examples
Model numeric aspects	**Numerics**: solver selections and settings, bugs **Simulators:** limitations of simulation platforms	**Reducible:** due to numerical approximations and software bugs	Lower imposed solver accuracy is chosen to allow for higher speed of calculations

of uncertainty but in most cases, this can be reduced by collecting additional data at the plant. Table 5.2 provides examples of input-related uncertainty and variability.

5.3.3.2 Model structure and parametric uncertainty

Each of the sub-models (influent model, activated sludge model, settling model, etc.) used within a plant-wide model contain sources of model structure uncertainty. Table 5.3 lists the sources of uncertainty linked to model structure and parameter values and classifies them as reducible or irreducible. Decreases in load due to industrial discharge reduction can be represented as scenarios by adjusting the historic influent loads. Uncertain kinetic parameter values can be described as probability functions with extreme values as bounds.

5.3.3.3 Model numerical uncertainty

Table 5.4 provides examples of uncertainty linked to numerical aspects of model implementation and classifies each source as reducible or irreducible (Benedetti *et al.*, 2012; Claeys *et al.*, 2010).

5.3.4 Prioritise and reduce sources of uncertainty

In this task, a reduced list of the sources of uncertainty that have the greatest impact on the specific project objective is compiled. To that list, the sources of uncertainty which will improve the general confidence in the model should be added. In this step the uncertainty matrix is also reduced.

Once the parameter ranges are defined, ***sensitivity analysis*** can help identify to which uncertainties the model outputs are most sensitive (Saltelli *et al.*, 2004).

The collection of additional data or the execution of experiments will reduce the spread of values of an uncertain parameter. Reducing uncertainty improves model predictive accuracy.

5.3.5 Describe sources of variability and uncertainty explicitly

5.3.5.1 Influent variability and generation of input time series

One of the major sources of uncertainty/variability with which both plant designers and operators have to deal is the dynamics of the influent. Typically, influent wastewater variability related to flow, temperature or water quality characteristics can be described using a time series when data are available.

In the absence of measurements of adequate frequency and duration, the appropriate variability can be introduced into the model by using an influent generator (Martin & Vanrolleghem, 2014) to randomly generate synthetic time (see Box 5.1). The synthetic time series must incorporate the underlying stochastic characteristics of the different variables and their correlations.

BOX 5.1 INFLUENT GENERATOR

An example of an influent generator, the one developed by Talebizadeh *et al.* (2016), uses two types of statistical models. One model for the synthetic generation of rainfall time series (a Markov chain-gamma distribution) and another for the time series describing the influent during dry weather flow (DWF) conditions (a multivariate autoregressive model with periodic terms). These two-time series (i.e., rainfall and influent in DWF conditions) serve as stochastic inputs to a conceptual model of the sewershed in order to generate the influent time series during both wet weather flow (WWF) and DWF conditions.

5.3.5.2 Parameter uncertainty

Uncertain parameters can be described in terms of probability density functions (PDFs). When site measurements are available the data can be used for the estimation of the parameters of the PDF. In the absence of historical data, subjective judgment (expert elicitation) can be used instead. Extreme parameter values can be designated as bounds. Any knowledge regarding the correlations among the different model parameters need to be also taken into account.

Different types of distribution functions (e.g., triangular or truncated normal distributions) can be used. In the absence of prior knowledge or site data the uncertainty can be characterised by assigning uniform distributions to the model parameters in order to avoid the under-estimation of uncertainty in model outputs (Freni & Mannina, 2010). Other than recognised and total ignorance, all uncertainties need to be described in such a way that they can be accounted for in the analysis, by propagating them through a model.

5.3.6 Model set-up and model structure uncertainty

In this task the selection of the category of models (i.e., for activated sludge selection between ASM1, ASM2d, …) and the complexity of the layout is decided. Model selection and layout complexity are usually dictated by the project objective, process configuration under investigation and the available data. The model variables, structure and the links between system components and processes contribute to model structure uncertainty.

Identifiability analysis can be used to 'expose inadequacies in the data or suggest improvements in the model structure' (Matott *et al.*, 2009). Identifiability can be defined as the situation where it is difficult to give a unique value to the model parameters. This is usually the result of the combination of (1) a model with a rather large number of parameters and (2) lack of sufficient data or lack of data of high quality (see Appendix A for more details).

Model structure uncertainty or uncertainty about expressing the relationship among the different variables of a system (Beck, 1987) as well as evaluation of different model structures for selecting the optimum model structure for dynamic simulation of WWTPs are not addressed in this methodology.

5.3.7 Propagation of uncertainty and variability using Monte Carlo simulation

Following model set-up, variability and uncertainty can be propagated through the model by running simulations with different influent time series and random samples from the joint distribution of model parameters selected as uncertain and quantified as described in Sections 5.3.3–5.3.5. This will generate different realisations of operational and effluent time series (different 'possible realities').

The simulation outputs are aggregated and evaluated after each run and output metrics are calculated. These metrics, such as the probability of compliance or non-compliance (PONC) provide quantitative estimates of the stochastic features of the system response (Talebizadeh, 2015). Figure 5.7 summarises the uncertainty propagation and effluent metric estimation procedure, and is further described in the sections below.

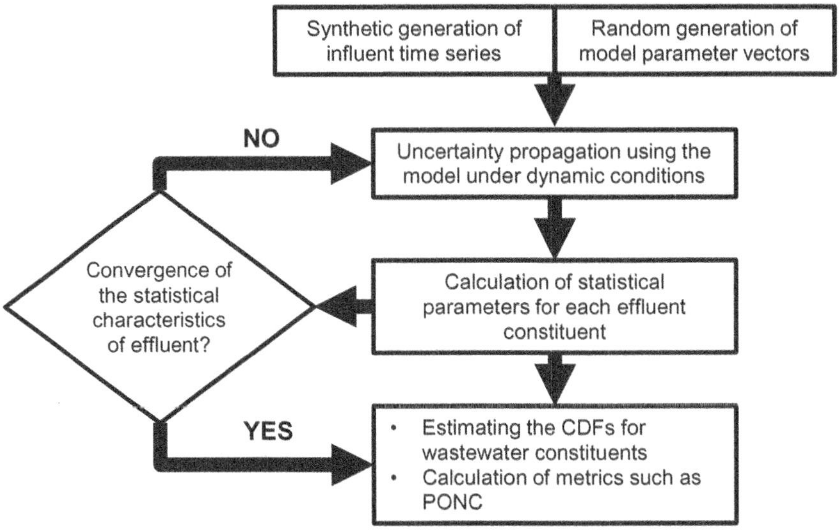

Figure 5.7 Uncertainty propagation and effluent metric estimation procedure (Talebizadeh, 2015).

5.3.7.1 Monte Carlo simulations

Monte Carlo simulation in combination with modelling is the most common method implemented for uncertainty propagation as discussed in Chapter 4 (see also Benedetti *et al.*, 2006; Bixio *et al.*, 2002; Huo *et al.*, 2006; Rousseau *et al.*, 2001; Sin *et al.*, 2009).

Depending on how uncertainty and variability are propagated through the model, different types of Monte Carlo methods can be applied. Monte Carlo simulation can be implemented using either a one-dimensional or a two-dimensional approach. To address short-comings in these two methods, Talebizadeh (2015) proposed a third approach called the pragmatic Monte Carlo method. All three of these approaches are explained below.

In each Monte Carlo run, a different vector of model parameters is used (Figure 5.8). Each vector is randomly sampled from the joint (the probability of two events occurring simultaneously) or marginal (the probability of an event irrespective of the outcome of another variable) distributions of the uncertain model parameters.

Two commonly used sampling methods used in the wastewater treatment field are the random sampling (RS) and Latin hypercube sampling (LHS) (Benedetti *et al.*, 2011; Stein, 1987). A short description of these two methods has been included in Appendix B.

Monte Carlo Convergence

Monte Carlo runs must continue until the statistical properties of the different effluent constituents such as average, standard deviation or a certain percentile, become stable (Benedetti *et al.*, 2011). For example, the fluctuation in the average, 5th, 50th, 95th percentiles can be evaluated at each Monte Carlo run, and simulations continued until the relative changes in the values of these four statistics drop below a pre-determined percentage (e.g., 1%). The effluent time series obtained may require some time-series aggregation prior to the convergence test.

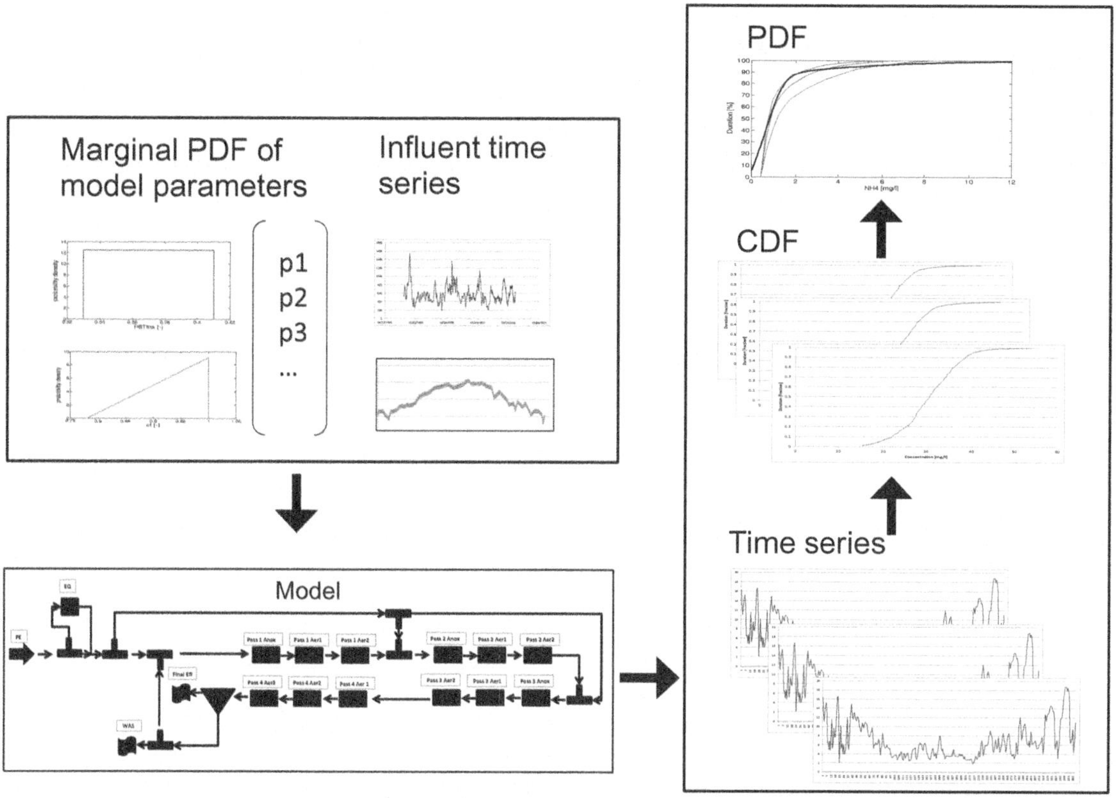

Figure 5.8 Graphic description of Monte Carlo simulation.

5.3.7.2 One-dimensional Monte Carlo simulation

In one-dimensional Monte Carlo simulation, no distinction is made between uncertainty and variability. In order to obtain the effluent distribution, the following steps are followed (Talebizadeh, 2015):

(1) Random generation of a vector of uncertain model parameters.
(2) Synthetic generation of a year-long influent time series (generated by an influent generator).
(3) Running the dynamic simulation of a design alternative using the parameter values in (1) and the influent time series in (2).
(4) Repeating 1–3 N times until the effluent distribution passes the specified convergence test.

Because uncertainty and variability are lumped together (uncertainty in model parameters and variability in the influent time series), a sample of uncertain model parameters (step (2) above) and influent time series (step (1) above) are input simultaneously to the dynamic model for each design alternative in each Monte Carlo run.

One of the main problems that can arise by combining uncertain and variable sources is that the information regarding the contribution of each source is lost and the result may become technically difficult to interpret (Wu & Tsang, 2004). If a one-dimensional Monte Carlo simulation is to be used correctly for uncertainty analysis, the effect of either uncertainty or variability on the key outcomes of a design (e.g., tank volumes, effluent quality), must be negligible (Merz & Thieken, 2005).

5.3.7.3 Two-dimensional Monte Carlo simulation

Two-dimensional Monte Carlo simulation is comprised of two loops. A variability loop is nested inside an uncertainty loop. This allows variability and uncertainty to be considered separately (Frey & Rhodes, 1996). For a given design alternative, the uncertainty loop will be executed *P* times. Within each iteration of the uncertainty loop, the variability loop will be executed *N* times. Neither the value for *P* or *N* is known at the beginning of this analysis. Instead, at the end of each iteration of either loop, the effluent output series that is generated is subjected to a statistical analysis to determine if pre-determined convergence criteria have been satisfied. If the answer is yes for the variability loop, then a new iteration for the uncertainty loop is begun. If yes for the uncertainty loop, then the uncertainty analysis is considered completed.

The execution of a two-dimensional Monte Carlo simulation can be summarised as follows (Talebizadeh, 2015):

(1) Begin an iteration of the uncertainty loop by randomly sampling from the probability distribution of each uncertain parameter. This generates values for a vector of uncertain parameters that will be used throughout Step 2.
(2) Begin an iteration of the variability loop.
    (a) Using an influent generator, synthetically generate a year-long influent time series.
    (b) Simulate effluent quality time series for the influent time series generated in Step 2a and for the randomly chosen uncertain parameters from Step 1.
    (c) Calculate statistics for the effluent time series from all runs completed for Step 2 to determine if the a priori established convergence criteria for the effluent distribution for this variability dimension have been met.
    (d) If convergence has been achieved (*N* simulations), continue to Step 3. If not, repeat Steps 2a through 2c.
(3) Go to Step 1 for the next iteration of the uncertainty loop by randomly sampling from the probability distribution of each uncertain parameter. Follow Steps 2a through 2d.
(4) Evaluate data to determine whether the convergence criteria for the uncertainty dimension have been satisfied (*P* simulations).
    (a) If convergence is not satisfied, return to Step 1;
    (b) If convergence is satisfied end the uncertainty loop.

The result of the two-dimensional Monte Carlo simulation is a cloud of CDF distributions (rather than a single CDF for one-dimensional Monte Carlo). Each CDF represents plant performance with different levels of conservatism (Figure 5.9).

The two-dimensional Monte Carlo simulation is very computationally expensive (Hoffman & Hammonds, 1994), however, if necessary, more efficient computing methods like cluster computing can reduce overall simulation time (Benedetti *et al.*, 2008; Claeys *et al.*, 2006).

5.3.7.4 Pragmatic Monte Carlo method

Talebizadeh (2015) proposed a novel approach that addresses the shortcomings of the one-dimensional Monte Carlo simulation method and has a substantially lower computational load compared to the two-dimensional Monte Carlo method.

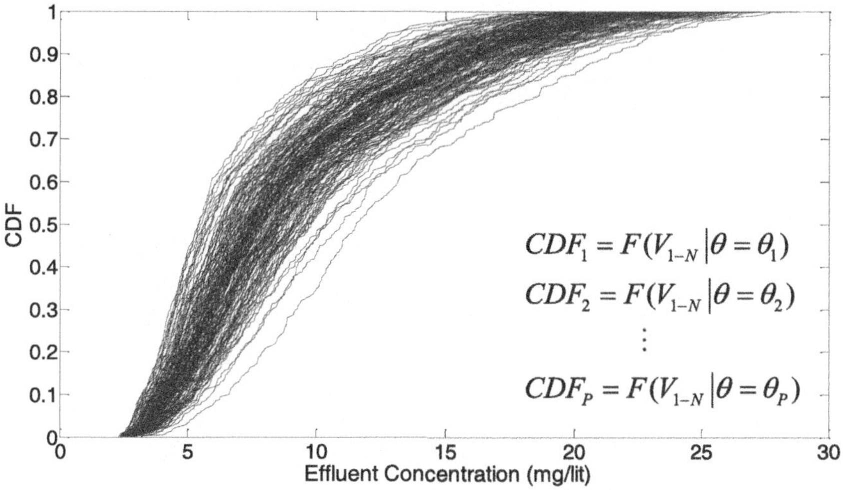

Figure 5.9 A cloud of effluent CDFs resulting from a two-dimensional Monte Carlo simulation. (Talebizadeh, 2015). The graph shows $N \times P$ number of CDFs. Each CDF has been generated by fixing the uncertain model parameters at a particular vector of parameter values and running the dynamic model of the plant under N different influent time series.

In the proposed method, named the pragmatic Monte Carlo method, uncertainty is evaluated at a small number of particular vectors of model parameters. One vector of model parameters could correspond to values obtained from previous studies on the plant, a calibrated model or expert knowledge. Another vector could represent a 'worst-case' scenario that would result in a conservative effluent CDF. A third vector could represent parameter values representing a very aggressive design. For details on the proposed method and the procedure for identifying a 'worst-case' vector of model parameters, see Talebizadeh (2015).

Once the vectors of model parameters are selected, one-dimensional Monte Carlo simulations can be run for each of the selected vectors with different influent time series (generated by an influent generator).

5.3.7.5 Effluent constituents cumulative distribution generation

As stated previously, at the end of the Monte Carlo simulations the CDF of each of the effluent constituents of interest is generated. It must be noted that the simulated effluent time series may require aggregation before deriving any statistics to check the convergence of the Monte Carlo runs or to compare with effluent permits. Time-series aggregation is necessary when the temporal resolution of the model output (simulated effluent time series) is different than the one by which compliance to a specific effluent standard is measured. For example, if the simulated effluent time series has a temporal resolution of 15 min and the compliance to effluent standards is measured based on flow-proportional daily-average concentration values, then the effluent time series (with 15 min temporal resolution) needs to be aggregated to daily effluent time series (Figure 5.10). Once the simulated effluent time series is aggregated, the convergence of effluent distributions can be evaluated for the three types of uncertainty propagation methods explained in Section 5.3.7.1.

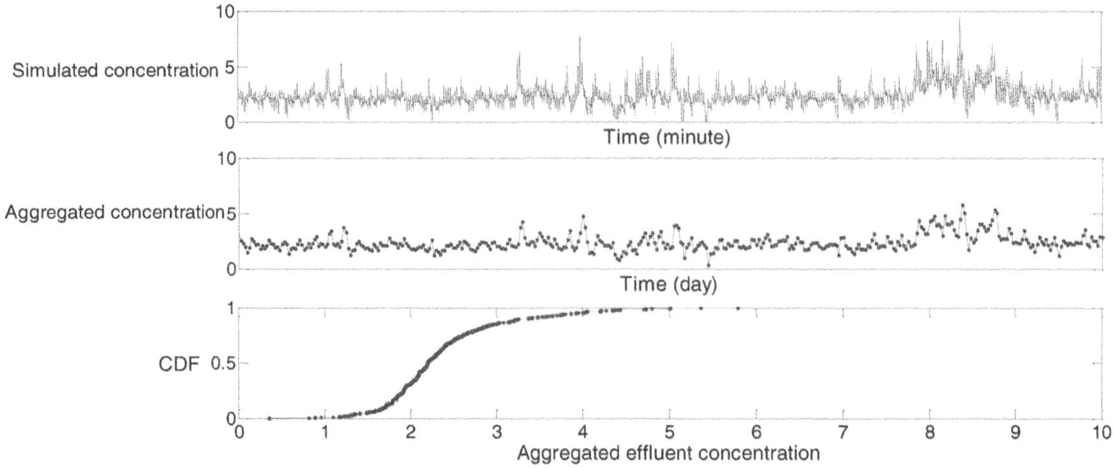

Figure 5.10 Aggregation of the simulated concentration time series ($\Delta t = 15$ min) into daily values and calculation of PONC using the empirical CDF (bottom) (Talebizadeh, 2015).

5.3.8 Synthesise evaluation metrics (output analysis)

Once all Monte Carlo runs are completed, evaluation metrics for each design alternative can be compiled and compared. These metrics will quantify the level of risk or conservatism of each design under investigation. A variety of metrics can be devised however, the most important two are:

- The probability of compliance or PONC which is directly linked to the effluent permit under which the design will be operating.
- The total cost of each design alternative.

These two quantitative criteria are used (in conjunction with other qualitative criteria) for the selection of the optimum design alternative. They are described in more detail below.

5.3.8.1 Calculation of PONC

Figure 5.11 shows the CDF for an example effluent constituent (of a specific design alternative), generated by means of the uncertainty analysis discussed in the previous sections. This CDF can be used to estimate a PONC for that constituent. Taking the example of an effluent limit for the constituent of 1 mg/L, the CDF indicates that the design alternative would produce an effluent concentration for that constituent that is 1 mg/L or less for 92% of the compliance periods. For 8% of the compliance periods, the 1 mg/L effluent limit would be exceeded resulting in a non-compliance event.

The PONC value represents the expected ratio of non-compliance events to the total number of events. For example, if the effluent permit is based on daily average concentration values, then each day constitutes either a compliance or non-compliance event. In the example shown in Figure 5.11, the expected number of days of non-compliance can be calculated by multiplying the value of PONC corresponding to the effluent permit (in this case 1 mg/L) with the total number of days in a year (i.e., $0.08 \times 365 = 29$). It must be noted that the number of days of non-compliance may vary significantly for different years depending on the different realisations of the influent time series as well as the set of model parameters used in the particular Monte Carlo run.

Depending on effluent compliance laws, in certain jurisdictions, some non-compliance events are allowed. In such cases, knowing the probability of a certain number of events occurring in a year will be of interest. Calculating the probability of a specific number of days of non-compliance requires the estimation of a discrete probability distribution (a probability distribution that can take on a countable number of values) from the Monte Carlo simulation outputs. This CDF is created using the number of days of non-compliance at a specific effluent target, obtained after each dynamic simulation run. The total number of Monte Carlo simulations will be determined as described in Section 5.3.7.1. Figure 5.12 illustrates a CDF describing the number of non-compliance events that may occur in a year. For example, the probability of having 10 or fewer non-compliance events in a year is 0.87. In other words, the probability of having more than 10 non-compliance events in a year equals 0.13.

The application of Monte Carlo simulation in conjunction with process-based models for estimating the PONC in WRRF systems has already been reported in several studies (Benedetti *et al.*, 2006; Bixio *et al.*, 2002; Cierkens *et al.*, 2011; Martin *et al.*, 2012; Rousseau *et al.*, 2001).

5.3.8.2 Calculation of total cost

Following the calculation of the PONC values, the second metric of interest to be calculated is the total cost (capital and operational) associated with each design alternative. The cost of a WRRF can be calculated using cost functions (Benedetti *et al.*, 2006; Bode & Lemmel, 2001; Gillot *et al.*, 1999) which are typically region specific or engineering company specific.

The calculated total cost of the design alternatives under investigation can be plotted against their corresponding PONC values to help identify the design alternatives with the most appropriate cost and PONC combination.

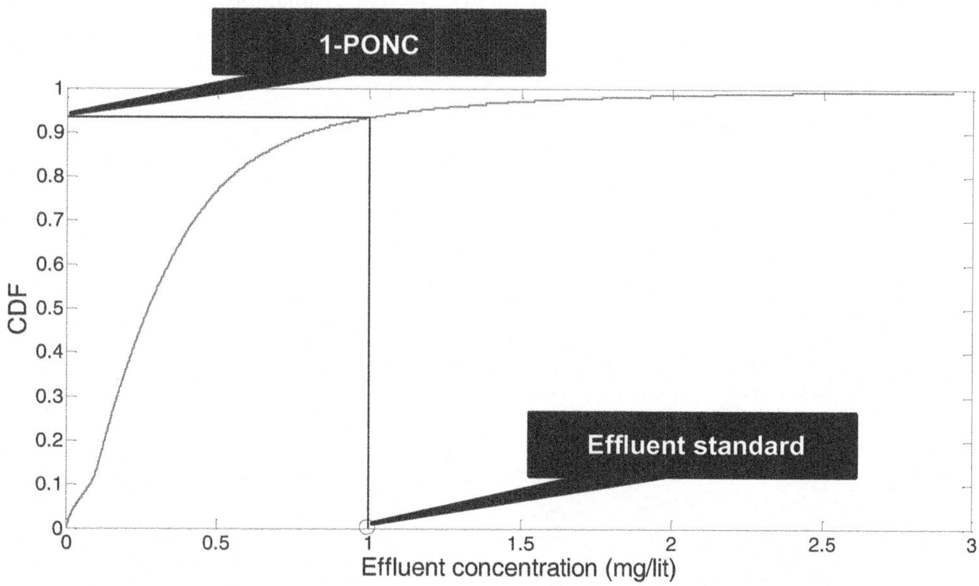

Figure 5.11 Calculation of PONC corresponding to an effluent standard (Talebizadeh, 2015).

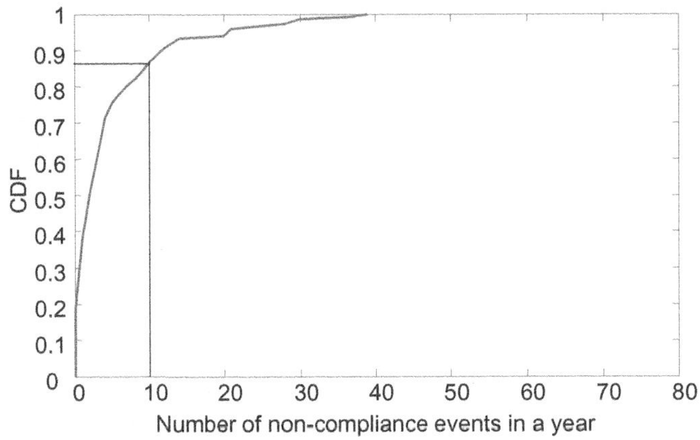

Figure 5.12 CDF for the number of non-compliance events in a year based on *N* years of simulation (Talebizadeh, 2015).

5.3.9 Communicate results

The tools and methods used for the communication of model results following an uncertainty analysis will depend on who is looking at the model outputs. There is currently no established best practice for communication and visualisation of model results and uncertainty and probabilities can be difficult to communicate effectively to some stakeholders.

The most common graphical displays of probabilistic information are probability density functions (PDFs), cumulative density functions, and box-and-whisker plots.

Graphics that list the sources of uncertainty and describe the impact of each source of uncertainty on key outputs (e.g., reactor size, cost) will be an important aid to decision making.

Selecting the optimal design can be aided by multi-objective evaluation/optimisation methods. The selection can be narrowed down by a Pareto front. An example of such multi-criteria analysis comparing different WRRF configurations under uncertainty is shown in Figure 5.13 (Benedetti *et al.*, 2008). The

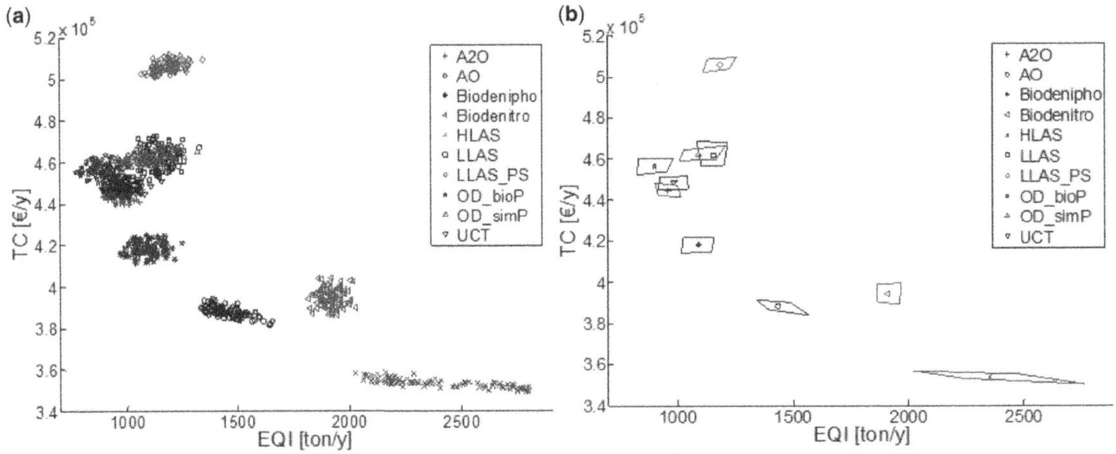

Figure 5.13 Pareto front of 10 30 000 PE configurations studied for total cost (TC) and effluent quality (EQI) (Benedetti *et al.*, 2008).

figure shows a comparison of the simulated annual average effluent quality index (EQI) and the total cost (TC) for 10 treatment plant configurations. The figure on the left (A) plots a cloud of 100 dots for each of the 10 configurations compared, with each dot representing EQI and TC for one MC simulation. In the figure on the right (B), each cloud is summarised by a polygon joining the 5th and 95th percentiles for the EQI and TC and by a marker for the 50th percentile. The larger the projection of a configuration's polygon on an axis is, the larger the uncertainty of that configuration for the variable plotted on that axis.

Such outputs encourage decision makers to make trade-offs explicit by using multi-criteria methods and make the decision more transparent to stakeholders.

5.4 SUMMARY

In this chapter the outline of a probabilistic design method for the design of WRRFs was presented. The method includes identifying the relevant sources of uncertainties and characterising them where possible with probability distribution functions (PDFs).

The sources of uncertainty include numerical uncertainties resulting from the selection of a numerical solver, uncertainties stemming from the selection of model structure, inputs and parameters and uncertainties linked to the equipment selection and operational procedures that result in the desired plant reliability.

PONC is proposed as a metric for the stochastic evaluation of the response of a specific design. The PONC is calculated using a dynamic model of the plant and Monte Carlo simulation. Calculating PONC as a quantitative measure of safety for each design alternative helps designers better understand and compare the performance of different design alternatives. Even in projects in which the sizing of a WRRF should be consistent with a specific design guideline, the proposed probabilistic design can be used as a tool for selecting proper values for safety factors and other inputs that are required for dimensioning the different units of a WRRF.

It should be noted that the proposed methodology does not cover all sources of uncertainty. For example, sources of uncertainty such as plant failures due to equipment (e.g., pumps or sensors) malfunction were not considered. Considering the impact of equipment failure on the PONC requires including their performance in the model and implementing reliability analysis. This was outside the scope of this study. Equipment failure models exist in the literature but have not yet been implemented in commercial simulators.

REFERENCES

ATV. (2000). Standard ATV-DVWK-A 131E, Dimensioning of Single-Stage Activated Sludge Plants. German Association for Water, Wastewater and Waste, Hennef, Germany, p. 57.

Beck M. B. (1987). Water quality modelling: a review of the analysis of uncertainty. *Water Resources Research*, **23**(8), 1393–1442.

Belia E., Amerlinck Y., Benedetti L., Johnson B., Sin G., Vanrolleghem P. A., Gerney K. V., Gillot S., Neumann M. B., Rieger L., Shaw A. and Villez K. (2009). Wastewater treatment modelling: dealing with uncertainties. *Water Science and Technology*, **60**(8), 1929–1941.

Benedetti L., Bixio D. and Vanrolleghem P. (2006). Benchmarking of WWTP design by assessing costs, effluent quality and process variability. *Water Science and Technology*, **54**(10), 95–102.

Benedetti L., Bixio D., Claeys F. and Vanrolleghem P. A. (2008). Tools to support a model-based methodology for emission/immission and benefit/cost/risk analysis of wastewater systems that considers uncertainty. *Environmental Modelling & Software*, **23**(8), 1082–1091.

Benedetti L., Claeys F., Nopens I. and Vanrolleghem P. A. (2011). Assessing the convergence of LHS Monte Carlo simulations of wastewater treatment models. *Water Science and Technology*, **63**(10), 2219–2224.

Benedetti L., Batstone D., De Baets B., Nopens I. and Vanrolleghem P. A. (2012). Uncertainty analysis of WWTP control strategies made feasible. *Water Quality Research Journal of Canada*, **47**(1), 14–29.

Bixio D., Parmentier G., Rousseau D., Verdonck F., Meirlaen J., Vanrolleghem P. A. and Thoeye C. (2002). A quantitative risk analysis tool for design/simulation of wastewater treatment plants. *Water Science and Technology*, **46**(4), 301–307.

Bode H. and Lemmel P. (2001). International product cost comparison in the field of water management. *Water Science and Technology*, **44**(2–3), 85–93.

Cierkens K., Belia E., Benedett L., Weijers S., Flameling T., De Baets B. and Nopens I. (2011). Comparison of implicit and explicit uncertainty analyses in the design of a WWTP upgrade. Proceedings of Watermatex 2011, 8th IWA Symposium on Systems Analysis and Integrated Assessment, 20–22 June 2011, San Sebastiàn, Spain.

Claeys F., Chtepen M., Benedetti L., Dhoedt B. and Vanrolleghem P. A. (2006). Distributed virtual experiments in water quality management. *Water Science and Technology*, **53**(1), 297–305.

Claeys P., Van Griensven A., Benedetti L., De Baets B. and Vanrolleghem P. A. (2010). On numerical solver selection and related uncertainty terminology. *Journal of Hydroinformatics*, **12**, 241–250.

Frey H. C. and Rhodes D. S. (1996). Characterising, simulating, and analysing variability and uncertainty: an illustration of methods using an air toxics emissions example. *Human and Ecological Risk Assessment*, **2**(4), 762–797.

Freni G. and Mannina G. (2010). Bayesian approach for uncertainty quantification in water quality modelling: the influence of prior distribution. *Journal of Hydrology*, **392**(1–2), 31–39.

Gillot S., De Clercq B., Defour D., Simoens F., Gernaey K. V. and Vanrolleghem P. A. (1999). Optimisation of wastewater treatment plant design and operation using simulation and cost analysis. Proceedings 72nd WEFTEC Conference, 9–13 October 1999, New Orleans, USA.

Hartigan J. and Wong M. (1979). Algorithm AS 136: a K-means clustering algorithm. *Journal of the Royal Statistical Society, Series C (Applied Statistics)*, **28**(1), 100–108.

Hoffman F. O. and Hammonds J. S. (1994). Propagation of uncertainty in risk assessments: the need to distinguish between uncertainty due to lack of knowledge and uncertainty due to variability. *Risk Analysis*, **14**(5), 707–712.

Huo J., Jiang Y., Seaver W. L., Robinson R. B. and Cox C. D. (2006). Statistically based design of wastewater treatment plants (WWTPs) using Monte Carlo simulation of Activated Sludge Model No.1 (ASM1). Proceedings of the 2006 World Environmental and Water Resources Congress, 21–25 May 2006, Omaha, Nebraska, USA.

Iman R. L. and Conover W. J. (1982). A distribution-free approach to inducing rank order correlation among input variables. *Communications in Statistics – Simulation and Computation*, **11**, 311–334.

Jakeman A. J., Letcher R. A. and Norton J. P. (2006). Ten iterative steps in development and evaluation of environmental models. *Environmental Modelling & Software*, **21**, 602–614.

Martin C. and Vanrolleghem P. A. (2014). Analysing, completing, and generating influent data for WWTP modelling: a critical review. *Environmental Modelling & Software*, **60**, 188–201.

Martin C., Neumann M. B., Altimir J. and Vanrolleghem P. A. (2012). A tool for optimum design of WWTPs under uncertainty: estimating the probability of compliance. Proceedings International Congress on Environmental Modelling and Software (iEMSs2012), 1–5 July 2012, Leipzig, Germany.

Matott L. S., Babendreier J. E. and Purucker S. T. (2009). Evaluating uncertainty in integrated environmental models: a review of concepts and tools. *Water Resources Research*, **45**, W06421, doi:10.1029/2008WR007301

Metcalf & Eddy Inc., Tchobanoglous G., Burton F. L., Tsuchihashi R. and Stensel D. (2013). Wastewater engineering: Treatment and resource recovery (5th ed.). McGraw-Hill Professional.

Merz B. and Thieken A. H. (2005). Separating natural and epistemic uncertainty in flood frequency analysis. *Journal of Hydrology*, **309**(1), 114–132.

Refsgaard J. C., van der Sluijs J. P., Højberg A. L. and Vanrolleghem P. A. (2007). Uncertainty in the environmental modelling process – a framework and guidance. *Environmental Modelling & Software*, **22**(11), 1543–1556.

Rieger L., Gillot S., Langergraber G., Ohtsuki T., Shaw A., Takacs I. and Winkler S. (2013). Guidelines for Using Activated Sludge Models, Scientific and Technical Report No.22, IWA Publishing, London, UK.

Rousseau D., Verdonck F., Moerman O., Carrette R., Thoeye C., Meirlaen J. and Vanrolleghem P. A. (2001). Development of a risk assessment-based technique for design/retrofitting of WWTPs. *Water Science and Technology*, **43**(7), 287–294.

Saltelli A., Chan K. and Scott E. (2004). Sensitivity Analysis. Wiley, New York, NY, USA.

Sin G., Gernaey K. V., Neumann M. B., van Loosdrecht M. and Gujer W. (2009). Uncertainty analysis in WWTP model applications: a critical discussion using an example from design. *Water Research*, **43**(11), 2894–2906.

Stein M. L. (1987). Large sample properties of simulations using Latin hypercube sampling. *Technometrics*, **29**, 143–151.

Talebizadeh M. (2015). Probabilistic Design of Wastewater Treatment Plants. PhD thesis, Département de génie civil et de génie des eaux, Université Laval, Québec, QC, Canada. https://corpus.ulaval.ca/jspui/handle/20.500.11794/26196 (accessed 2021).

Talebizadeh M., Belia E. and Vanrolleghem P. A. (2016). Influent generator for probabilistic modelling of nutrient removal wastewater treatment plants. *Environmental Modelling & Software*, **77**, 32–49.

Walker W. E., Harremoes P., Rotmans J., van der Sluijs J. P., van Asselt M. B. A., Janssen P. and Krayer von Krauss M. P. (2003). Defining uncertainty: a conceptual basis for uncertainty management in model-based decision support. *Integrated Assessment*, **4**(1), 5–17.

Wu F. and Tsang Y. (2004). Second-order Monte Carlo uncertainty/variability analysis using correlated model parameters: application to salmonid embryo survival risk assessment. *Ecological Modelling*, **177**, 393–414.

Chapter 6

Case studies

6.1 INTRODUCTION

This chapter presents two examples that illustrate the application of uncertainty analysis, in combination with process models, to inform decisions on treatment plant design and operations. These case studies illustrate how these tools can be utilised to quantitatively define the risks and opportunities in different design and operational decisions, and thus how a utility might select the appropriate levels of cost and risk. The examples include a steady state and a dynamic application.

6.2 STEADY-STATE UNCERTAINTY ANALYSIS EXAMPLE: OPERATION OF THE DURHAM WRRF

6.2.1 Project objectives

In this example, Clean Water Services (CWS) (Tigard, Oregon, USA) was exploring how to best operate their Durham Advanced Wastewater Treatment Facility (Figure 6.1) in anticipation that the local regulatory authority would require it to nitrify year around (Menniti *et al.*, 2014). Their permit at the time only required nitrification during the summer (dry) season. In reality, the dry season extends to a large part of the year and during this period the plant contributes a significant fraction of the river flow. Nitrification is needed to dilute the plant's effluent ammonia. However, the expected winter (wet) season effluent permit ammonia would be based in part on the receiving river flow, with lower river flows requiring higher levels of nitrification.

Operations staff wished to understand what operating sludge age they would need to target in the winter that would allow them to reliably achieve the required winter effluent ammonia targets.

 Uncertainty in Wastewater Treatment Design and Operation: Addressing Current Practices and Future Directions
Editor(s): Evangelia Belia, Lorenzo Benedetti, Bruce Johnson, Sudhir Murthy, Marc Neumann, Peter Vanrolleghem and Stefan Weijers
doi: 10.2166/9781780401034_0095

Figure 6.1 Durham advanced wastewater treatment facility.

6.2.2 Conventional design approach using safety factors

As discussed previously, design safety factors are normally based on industry experience for developing a design robust enough to accommodate: (1) future variability (2) uncertainty in operating conditions and (3) uncertainty in effluent requirements. The nitrification safety factor (NSF) is a widely applied heuristic ('rule of thumb') used to estimate the design sludge retention time (SRT) of a nitrifying activated sludge system (Scheible *et al*, 1993). The safety factor lumps together various performance-related uncertainties including vulnerability to inhibitory substances in the influent wastewater, pH swings, and difficulties in maintaining adequate dissolved oxygen.

The EPA Nitrogen Control Manual notes that safety factors are ultimately expressions of design confidence. For example, in the 1993 USEPA Manual on Nitrogen Control (Scheible *et al.*, 1993), as part of a design approach for a nitrifying suspended growth system the following is mentioned: 'the anticipated variations in process conditions and the uncertainty in the kinetic coefficients warrant a safety factor of 2.0' (Scheible *et al*, 1993). An overly conservative choice of a safety factor can lead to an unnecessarily expensive design. Conversely, a safety factor that is too low can lead to a plant that frequently fails to achieve its effluent ammonia target. The 1993 USEPA (Scheible *et al*, 1993) document defines the minimum sludge age (SRT) as the SRT at which nitrifiers are just about to wash out of the system. The equation they provide for the washout SRT (SRT_{MIN}, for pH values <7.2) is given in equation 6.1. The NSF is defined in equation 6.2.

$$SRT_{\mathrm{MIN}} = \frac{1}{\mu_{\max} \times \theta_{\mu,\max}^{(T-20)} \times \left(\dfrac{DO}{DO + K_{OA}}\right) \times [1 - 0.833 \times (7.2 - pH)] - b \times \theta_b^{(T-20)}} \tag{6.1}$$

where:

SRT_{MIN} = washout sludge age for nitrifiers (days)
$\mu_{\max}$ = maximum specific growth rate at 20°C (1/d)
$\theta_{\mu,\max}$ = maximum specific growth rate temperature adjustment

DO = dissolved oxygen concentration (mg/L)
K_{OA} = oxygen half-saturation value for autotrophs (mg/L)
b = autotrophic decay rate (1/d)
θ_b = decay rate temperature adjustment
T = temperature in °C

$$NSF = \frac{SRT_{\text{Aerobic}}}{SRT_{\text{MIN}}} \tag{6.2}$$

where:

NSF = nitrification safety factor
SRT_{Aerobic} = actual (or design) operating SRT

6.2.3 Probabilistic design approach

CWS chose to use a probabilistic approach to determine a suitable NSF for wet weather operations. The approach used is described below.

Firstly (step 1), the anticipated wet weather ammonia effluent requirements were determined from an analysis performed by CWS based on calculations of ammonia toxicity in the river. Ammonia toxicity is based on river flows, pH, temperature and ammonia concentrations. These effluent ammonia requirements were expected to decrease as the flow in the Tualatin River (discharge location) decreased (i.e., increasing impact of plant effluent ammonia on lower river flows). Figure 6.2 shows these values. At river flows above 21.24 m^3/s (750 ft^3/s), the target effluent ammonia is actually higher than the plant effluent ammonia when not nitrifying, thus eliminating the need for nitrification.

Secondly (step 2), historical data were analysed, to estimate the frequency at which the combination of river flow and plant influent water temperature would require the plant to nitrify to meet the anticipated effluent ammonia limits. When river flow is high, there is greater capacity in stream to dilute the

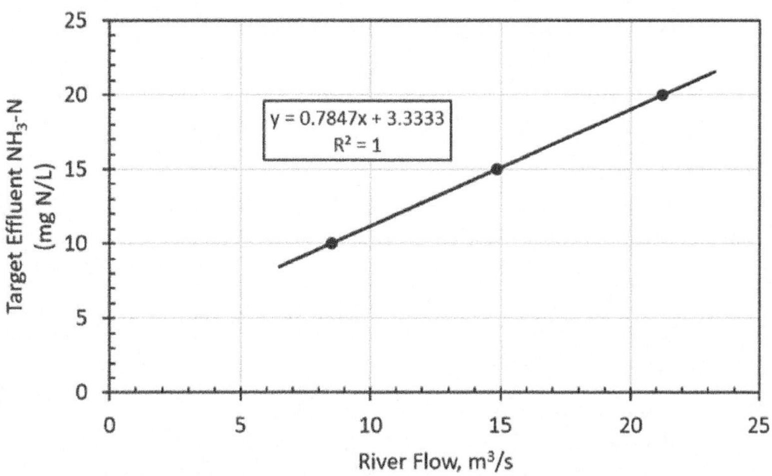

Figure 6.2 Target effluent ammonia requirements as a function of river flow.

ammonia received from the plant effluent. This lessens the extent of nitrification that must be accomplished within the plant. Influent temperature figures into the calculation because at higher temperature, nitrification proceeds at higher rates and sufficient nitrification can be achieved at lower aerobic SRT.

Thirdly (step 3), the results from step 2 were reviewed in order to select for planning purposes, the most appropriate target NSF. USEPA's Nitrification Safety Factor calculation (equation 6.2, Scheible *et al*, 1993) was used to determine the probability of achieving nitrification when river flows were low.

A Monte Carlo analysis was used for the probabilistic analysis. In a Monte Carlo analysis, the sources of uncertainty and variability in the parameters of a deterministic calculation are identified. For this project, the deterministic calculation is the NSF as described previously. The model input parameters considered variable or uncertain, defined by probability distribution functions (PDFs) and correlated against each other were:

- Wastewater influent temperature.
- Operating SRT. While determining this SRT was the goal of the work, the ability of operational staff to maintain this exactly is limited, therefore a normal distribution with a standard deviation of 1 day was set up around the target SRT.
- Nitrifier kinetic parameters. The maximum specific growth rate, μ_{max} (0.77 $\pm$ 5% 1/d), oxygen half-saturation coefficient, K_{OA} (0.05 $\pm$ 25% mg/l) and decay rate, b (0.5 $\pm$ 25% 1/d) were estimated through model calibration. These parameters were not measured directly and could vary over time. Therefore, their uncertainty and variability were accounted for in the probabilistic analysis using a uniform distribution following Sin *et al.* (2009).

Additionally, the river flow rate was correlated with the wastewater temperature. A probability density function is applied to those sources of uncertainty and variability to describe the range of possible parameter values. When possible, the probability density functions are fitted to historical data to ensure they describe actual conditions as accurately as possible. The influent temperature and operating SRT were fitted to historical data. These probability density functions are sampled hundreds or thousands of times to generate hundreds or thousands of possible parameter sets. The deterministic calculation is performed with each parameter set and the results are analysed to estimate the probability of different outcomes occurring. The impact of pH was also evaluated as described below.

Once the target NSF was determined, the sources of variability and uncertainty in the calculation of the NSF were identified, and this variability and uncertainty was quantified with PDFs.

Finally (step 4), to determine the design SRT, a Monte Carlo probability-based analysis was used. First, the reliability criteria were set with CWS's input to determine the acceptable level of risk assumed in the design. Then the design SRT was chosen for the planning alternatives to ensure the reliability criteria were satisfied. These risk criteria were:

- For a system pH of 7.2 (or for an assumption of no pH inhibition), the reliability criterion is that the NSF must be 1.3 or greater 95% of the time for the entire wet weather season.
- For a system pH of 7.0 (or for an assumption of nitrification inhibition due to low pH), the reliability criteria are that the NSF must be 1.3 or greater 95% of the time when the river flow is less than 21.24 m^3/s (750 ft^3/s), (the first benchmark river flow) and the NSF must be greater than 1.3, 75% of the time for the entire wet weather season.

Steps 1–4 described above did not involve a full plant simulation but, instead, used an NSF spreadsheet calculation with river input flows and temperatures (both wastewater and river) to determine the needed operating SRT, with the goal of running as low a SRT possible while minimising the probability of washing out nitrifiers during the winter period. For the final determination of the target minimum NSF, the SRT for nitrifier washout was also estimated with a steady state whole plant simulator based on IWA's ASM2d model and compared to the minimum SRT predicted from the EPA NSF equation.

For NSF calculations using the whole plant simulator, the model input SRT and the aerobic fraction of the plug flow basins were used to determine the NSF. The equation for SRT_{MIN} shown above was used to determine the washout SRT.

6.2.4 Results and discussion

The results from step 4 indicated that the nitrifiers start allowing significant ammonia in the effluent (begin to wash out of the system) just below an NSF of 1.3. This change in nitrification happens abruptly in the simulated plant, with the effluent ammonia increasing from around 1 mg-N/L to around 20 mg-N/L when the NSF decreased from 1.3 to 1.2. The fact that the model indicates loss of nitrification above an NSF of 1 indicates that the EPA approach and the simulation are not exactly in alignment, which is not surprising in light of the simple approach of the EPA equation vs. the ASM2d model. In reality, however, there is a wider band of operating conditions where nitrification is unstable, but the nitrifiers do not wash out. This is due to variability in SRT control, wastewater temperature, and other operating factors.

Imminent nitrifier washout in the operating plant was defined to occur when the plant effluent ammonia concentration increased above 1.0 mg-N/L. Nitrifier washout was predicted by the ASM2d-based simulator at an NSF of 1.3, indicating that the EPA NSF equation predicts nitrifier washout at a lower SRT than that predicted by the whole plant simulator. The same parameter values were used in NSF calculation as were used in the ASM2d model, accounting for the differences in the ASM2d death/regeneration approach.

The minimum target NSF of 1.3 was also confirmed with actual plant operating data (Figure 6.3), which further supports the idea that the EPA equation does not quite reflect actual kinetics. However, even in light

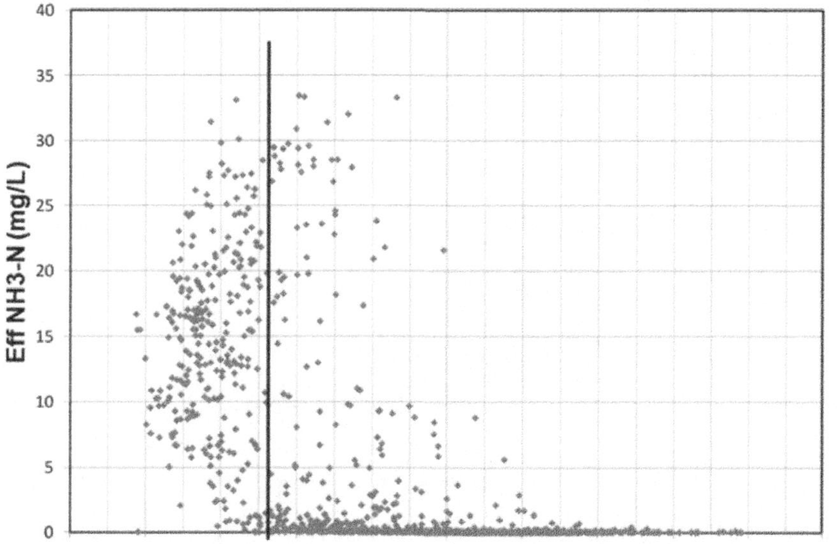

Figure 6.3 Durham AWTF operational results for determining the target NSF.

of this discrepancy, once adjusted for observed operations (i.e., 1.3 vs. 1.0 NSF), the NSF calculation provides a simple approach for understanding how close nitrification is to failure within an activated sludge system.

The wastewater influent temperature was found to be negatively correlated with the river flows. As the river flows went down (and decreased the target effluent ammonia, the wastewater influent temperature was found to increase. Figure 6.4 shows the measured values of river flow vs. the wastewater temperature as well as the equivalent sampled values from the probability model correlation that was set up between these two parameters.

In the absence of the probabilistic analysis, a 'rule of thumb' NSF, based on engineering and operations experience of 1.5 would have been applied to operation at the minimum week wastewater temperature, resulting in a design SRT of 8.5 days. This probabilistic analysis resulted in a design SRT of 8.0 days as this SRT satisfied the reliability criterion (NSF >1.3 95% of the time) as shown in Figure 6.5. In Figure 6.5, the 5% bar (i.e., 95% reliability) shows that at 8 days the NSF was at 1.34, while at 8.5 days it was at 1.44 (results not shown), which was unnecessarily high. The comparison of these two SRTs illustrates the level of unnecessary conservatism inherently saved by quantifying the uncertainties with probability analysis. The lower design SRT maximises existing infrastructure investment because it increases the rated capacity of the secondary process at a lower SRT while, very importantly, providing CWS with confidence that the system will perform under critical wet weather conditions.

The data set was also sorted so only parameter sets with river flows less than 21.24 m^3/s (the river flow triggering the need for nitrification) were evaluated (Figure 6.5 right). These results demonstrate that the NSF is greater than the minimum target value of 1.3, more than 99% of the time when nitrification is required, providing CWS further assurance (a level of conservatism) that the secondary process will be able to reliably nitrify under critical wet weather conditions.

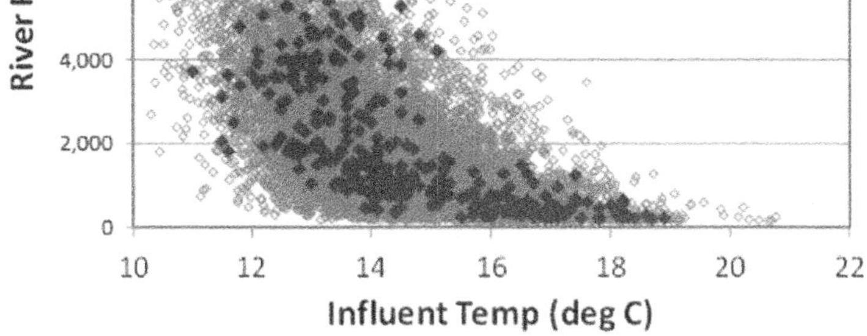

Figure 6.4 Correlation between influent temperature and river flow. Data in black are actual values, data in grey are results from PDF sampling.

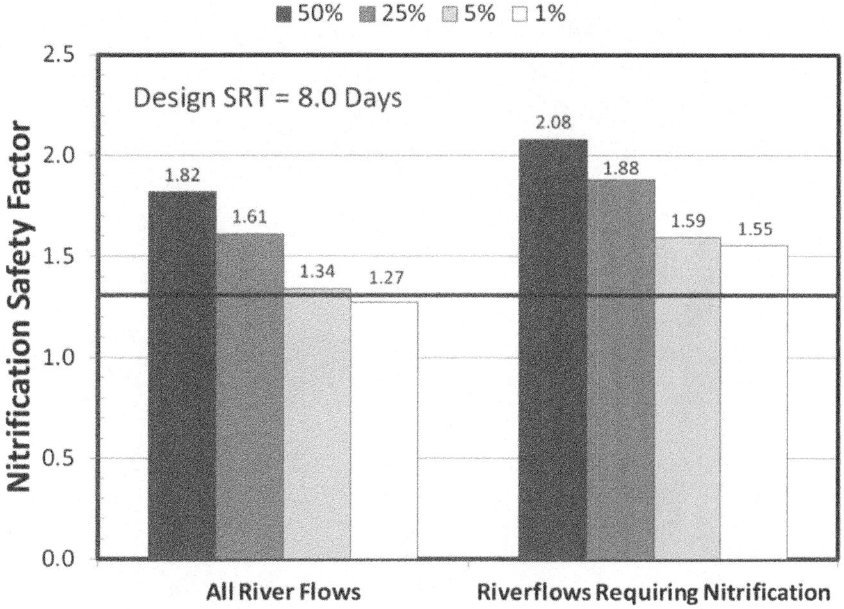

Figure 6.5 Durham AWTF NSF results showing the 50th, 25th, 5th, and 1st percentile NSF values for the chosen design SRT of 8.0 days and the results at 8.5 days SRT.

6.3 DYNAMIC UNCERTAINTY ANALYSIS EXAMPLE: DESIGN UPGRADE FOR THE EINDHOVEN WRRF

6.3.1 Project objectives

The Eindhoven WRRF (Figure 6.6) has a design capacity of 750 000 population equivalent (PE) and is the third largest WRRF in the Netherlands. Wastewater entering the plant is screened and de-gritted before going through primary treatment. The maximum design flow of the influent pumping station, preliminary treatment and primary clarifiers is 35 000 m^3/h (343 ft^3/s). However, the secondary treatment design flow is 26 250 m^3/h (258 ft^3/s). During high flow rates, excess flow is diverted to a storm storage tank. The biological treatment comprises three activated sludge tanks with anaerobic, anoxic, and aerated zones. Each activated sludge tank sends flow to four secondary clarifiers. The final effluent is discharged to the Dommel River. A detailed description of the plant can be found in Cierkens *et al.* (2012). Information on the plant effluent permit, as well as the basic characteristics of the connected sewershed can be found in Schilperoort (2011) and Belia *et al.* (2012).

Between 2003 and 2006 the Eindhoven WRRF underwent an upgrade to comply with new, more stringent, nutrient effluent limits (e.g., daily average flow proportional ammonia of 2 mg-N/L) and also to increase the hydraulic capacity of the secondary treatment from 20 000 to 26 250 m^3/hr (196–258 ft^3/s).

The objective of this study was to use the probabilistic design methodology presented in Chapter 5 and summarised in Figure 6.7 to determine the area and depth of the secondary clarifiers and the total bioreactor volume (aerobic, anaerobic and anoxic) for this upgrade.

Figure 6.6 Arial view of the Eindhoven WWPT.

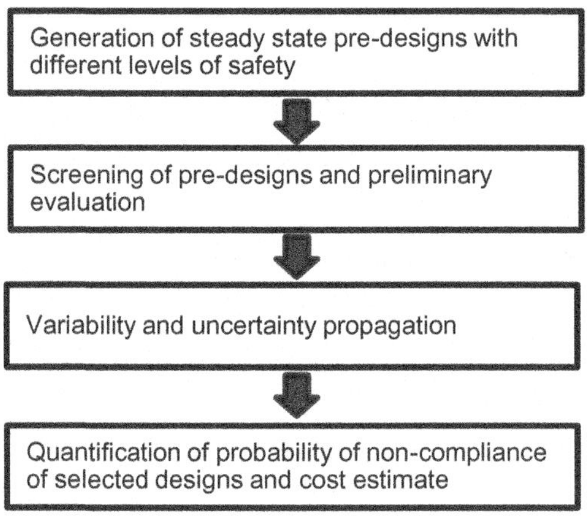

Figure 6.7 Proposed design methodology.

6.3.2 Generation and screening of steady-state pre-designs

The German ATV (2000) guidelines were used as a steady-state design tool for the generation of the pre-designs. A uniform uncertainty range was assigned to each input of the ATV design guideline parameters and the design outputs were generated by Monte Carlo simulation. The ranges of uncertainty were derived using the information obtained from the previous studies on the Eindhoven WRRF (Belia *et al.*, 2012; Schilperoort, 2011), ATV (2000) design guideline recommendations, effluent standards (imposed by regulations), and expert opinion. Table 6.1 shows a selection of the uncertain parameters for each major category. The complete list can be found in Talebizadeh (2015).

Table 6.1 Range of values assigned to the ATV design inputs (lower and upper limits of the uniform distribution).

ATV Design Inputs	Lower Limit	Upper Limit	Units
Influent constituents			
Primary effluent COD[1,2]	200	400	mg/L
Primary effluent nitrogen[1,2]	30	50	mgN/L
Maximum hourly wet weather flow rate as 2-h mean[1,2]	45 000	65 000	m^3/h
Inert particulate COD fraction of particulate COD[3]	0.2	0.35	%
Inorganic TSS fraction of total TSS[3]	0.2	0.3	%
Safety factors			
Safety factor for nitrification[3]	1.45	1.5	-
Safety factor applied to the effluent inorganic nitrogen[3]	0.6	0.8	-
Safety factor applied to the effluent phosphorous[3]	0.6	0.7	-
Operation parameters			
Minimum contact time in anaerobic tanks[1,2]	0.9	1.1	hr
Effluent concentrations			
Total nitrogen concentration in the effluent[4]	10	10	mgP/L
Phosphorous concentration in the effluent[4]	1	1	mgP/L

Notes: 1: Expert opinion, 2: Previous studies on the Eindhoven WRRF, 3: ATV design guideline, 4: Effluent standards.

Five thousand pre-designs were generated by random sampling of 5000 sets of ATV inputs. The ATV standards were applied to each of the 5000 sets to generate unit process dimensions for the aerobic, anoxic and anaerobic volume of the bioreactors and the area and depth of the secondary clarifiers. This resulted in 5000 alternative designs. The number of alternatives to be evaluated with a model run under dynamic conditions was reduced to the seven most representative ones by k-means clustering.

These seven design alternatives were representative of the overall design space of the outputs. The reduction of design alternatives for further evaluation through the k-clustering method keeps the computational load for the overall analysis at a manageable level.

The histograms on the diagonal panels in Figure 6.8 represent the distribution of design outputs (SST area and depth, total and anaerobic bioreactor volume) that were generated according to the ATV design guidelines. The red dots in the scatter plots of the other panels represent cluster centroids that were calculated using the *k*-means clustering method. For instance, the total activated sludge tank volume in the seven design alternatives varied between 71 000 and 107 000 m^3.

The selected design alternatives were further evaluated with a year-long influent time series, representing a typical year (Talebizadeh *et al.*, 2016). The simulation was performed using the ASM2d biological model (Henze *et al.*, 1999) and the Bürger *et al.* (2011) secondary settling model. For this set of simulations, the model parameters were given 'best estimate' values.

For each alternative a simulated effluent time series was generated. This effluent time series was processed to produce 24-hour mean values for COD, NH4-N, TSS and TN. Cumulative distribution functions (CDFs) were constructed for the four constituents. The objective of this step was to flag and eliminate alternatives that did not meet the desired performance criteria for a typical intra-annual variability. This step also identified alternatives that had very similar performance. Following this evaluation step five alternatives were selected for further analysis (Table 6.2).

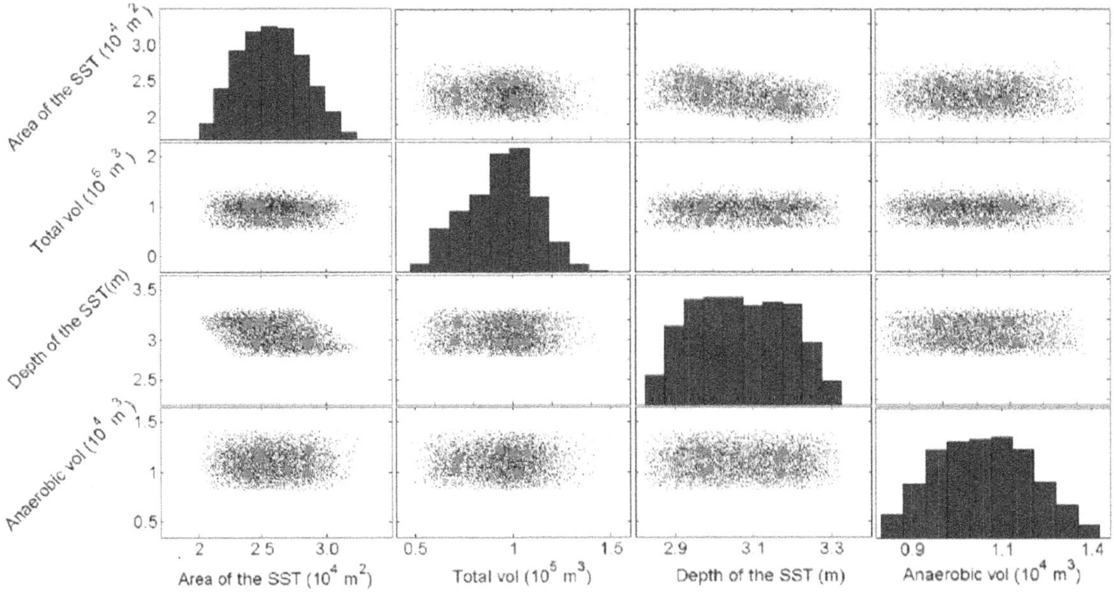

Figure 6.8 Distribution of the generated 5000 pre-designs and the centroids locations corresponding to the k-means clustering with seven centroids (i.e., the red dots). Total volume refers to total reactor volume.

Table 6.2 Dimensions of the design alternatives selected for further evaluation.

Design Alternatives	Total Reactor Volume (m^3)	Anaerobic Volume (m^2)	Depth of Secondary Clarifier (m)	Area of Secondary Clarifier (m^2)
Alt1	47 850	12 200	3.1	27 250
Alt2	59 400	11 100	3.0	25 250
Alt3	70 650	10 250	3.0	26 900
Alt4	106 650	11 850	3.0	24 600
Alt5	118 700	9500	3.1	26 250

6.3.3 Variability and uncertainty propagation

For the five alternatives selected for further evaluation, a pragmatic Monte Carlo simulation was performed, aimed at evaluating the impact of influent variability and parametric uncertainty on the performance of each design.

6.3.3.1 Influent variability

For the influent variability evaluation, as described in Section 5.3.5.1, *N* influent time series were generated by an influent generator calibrated using available weather, catchment and plant data. Details of the influent generator are described in Talebizadeh *et al.* (2016).

The influent data used for the calibration of the influent generator included flow, and sensor data for ammonia, soluble COD, total COD and TSS. Long-term daily rainfall data and also rainfall data with finer temporal resolution provided were used for estimating the parameters of the weather generator.

With the weather generator an indefinite number of rainfall intensity time series can be generated, each leading to one of the *N* influent time series.

6.3.3.2 Model parameter uncertainty

The uncertainty in the model parameters was characterised by assigning uniform distributions to uncertain model parameters. In this study all of the ASM2d model kinetic and stoichiometric parameters were considered uncertain. The lower and upper limits of the distribution of each parameter were calculated using a Nominal value (most likely value) multiplied by a percentage of the Nominal value as described in Brun *et al.* (2002). The Nominal values and uncertainty ranges were based on a combination of expert opinion, modelling experience and previous studies. Random sampling (RS) with no-correlation was selected for sampling from the distribution of uncertain model parameters. The choice of random sampling of model parameters was based on the study of Hauduc *et al.* (2011) in which no strong correlation was reported between the parameters of the ASM2d model.

The 'pragmatic' Monte Carlo method (for details see Chapter 5, Section 5.3.7.1) was implemented for the propagation of variability and uncertainty. The following sets of parameters were used:

(1) The lower limit of the uniform distributions
(2) The upper limit of the uniform distributions
(3) The Nominal set of model parameters
(4) The 'Worst Case' set of model parameters

Table 6.3 includes a sub-set of the model parameters considered uncertain. It shows the lower and upper limits of the uniform distributions used to describe the uncertainty surrounding the parameter values as well as the Nominal and 'Worst Case' values used in the uncertainty propagation simulations. The 'Worst Case' set of model parameters was selected to represent a very unfavourable condition for removal of ammonia and other parameters of interest. The 'Worst Case' set of model parameters corresponded to 95% confidence for the NH_4 effluent standards (i.e., 2 mg/l) and a higher than 95% for other pollutant concentrations. For a complete table see Talebizadeh (2015).

Table 6.3 Selected uncertain model parameters with their upper, lower limits, Nominal, and Worst-Case values.

Model Parameters	Lower Limit	Upper Limit	Nominal	Worst Case
Reference temperature of the activated sludge	20	20	20	20
Decay rate	0.075	0.225	0.15	0.224
Rate constant for lysis and decay	0.32	0.48	0.4	0.419
Hydrolysis rate constant	1.5	4.5	3	2.856
Maximum growth rate	0.8	1.2	1	0.862
Anoxic reduction factor for decay of autotrophs	0.165	0.495	0.33	0.386
Anoxic reduction factor for decay of heterotrophs	0.4	0.6	0.5	0.428
SVI	100	140	120	125

6.3.4 Quantification of probability of non-compliance (PONC)

From the simulated effluent time series of BOD, COD, and NH_4 (with 15-min temporal resolution) 24-h daily flow-proportional average concentrations were calculated. This matched the sampling frequency used for compliance. The plant also has discharge standards for TN and TSS that it must meet on an annual basis. Therefore, annual average concentrations for these water quality measures were calculated from the simulated effluent time series. Once the convergence of the effluent distributions was achieved, the CDFs of the different effluent constituents were derived and their corresponding PONC values were calculated.

For each design alternative, several PONC values were derived. Table 6.4 includes the PONC corresponding to the pragmatic Monte Carlo simulation at the Nominal and 'Worst Case' set of model parameters (refer to Chapter 5, Section 5.3.7.1 for discussion of the pragmatic Monte Carlo simulation). The PONC values calculated using the pragmatic Monte Carlo with model parameters set to the Nominal values correspond to the most likely behaviour of the plant. The PONCs calculated with model parameters set to the 'Worst Case' set of model parameters, correspond to a possible (but less likely compared to the Nominal set of model parameters) condition. As expected, the calculated PONC values are larger compared to the case of Nominal model parameters.

Table 6.4 also includes the expected number of days that the effluent ammonia and TN concentrations are expected to exceed the effluent standards. The metrics shown in Table 6.3 can be used to compare the behaviour of the design alternatives. As expected, an increase in bioreactor volume results in a reduction of PONC.

To better explore the relationship between the total volume of the bioreactors and the PONC, the PONC values for NH_4 of each design alternative were plotted against the total bioreactor volume (Figure 6.9).

The NH_4 PONC values for all of the design alternatives, calculated using the pragmatic Monte Carlo simulation at the Nominal set of model parameters are below 5%. However, the NH_4 PONC values for Alt1, Alt2, and Alt3 calculated at the 'Worst Case' set of model parameters (i.e., corresponding to a possible but conservative set of model parameters) are very high (i.e., 86.4, 78.7, and 29 expected days of non-compliance in a year, respectively), which may render them unacceptable due to their poor expected performance in NH_4 removal. In contrast to alternatives Alt1, Alt2 and Alt3, alternatives Alt4 and Alt5 have near zero PONCs at the Nominal set of model parameters and small values at the 'Worst

Table 6.4 PONC values for different design alternatives calculated using the pragmatic Monte Carlo simulation for two sets of model parameters ('Nominal' and 'Worst Case') (Talebizadeh, 2015).

Alternatives		**Alt1**	**Alt2**	**Alt3**	**Alt4**	**Alt5**
Total bioreactor volume		47 850	59 400	70 650	106 650	118 700
Pragmatic Monte Carlo simulation with Nominal parameter set						
NH_4	PONC	0.04	0.02	0.01	0.003	0.001
	Days[1]	15.1	6.2	3.4	1	0.2
TN	PONC	0.08	0.023	0	0	0
	Per cent[2]	8.00%	2.20%	0.00%	0.00%	0.00%
Pragmatic Monte Carlo simulation with Worst-Case parameter set						
NH_4	PONC	0.24	0.13	0.08	0.03	0.01
	Days[1]	86.4	48.7	29	7.8	5.1
TN	PONC	0.8	0.52	0.4	0.02	0.04
	Per cent[2]	80%	52%	40%	4%	2%

[1]Expected number of days with non-compliance event in a year (i.e., PONC $\times$ 365).
[2]Expected percentage of years with non-compliance events.

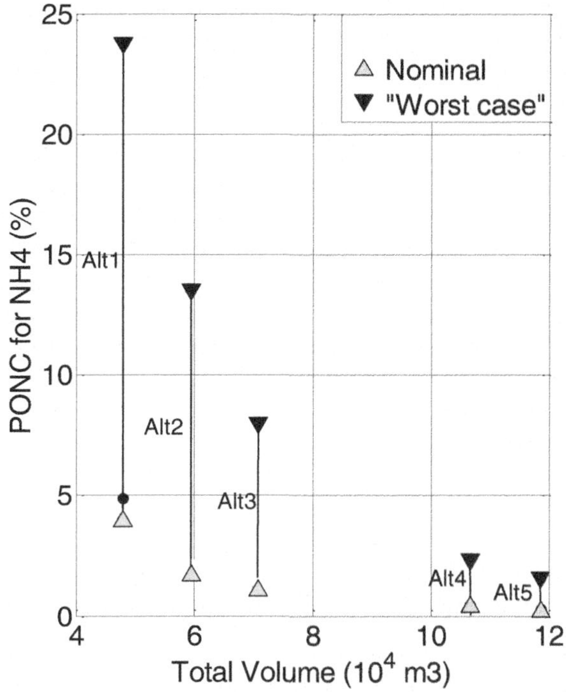

Figure 6.9 Relationship between PONC values (calculated using the pragmatic Monte Carlo simulation for the Nominal and 'Worst-Case' parameter sets) and the total bioreactor volume of the five design alternatives (Talebizadeh, 2015).

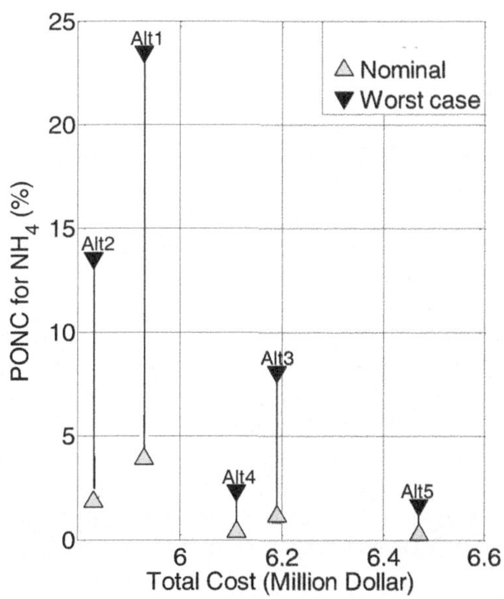

Figure 6.10 Relationship between PONC values (calculated using the pragmatic Monte Carlo simulation for the Nominal and 'Worst-Case' parameter sets) and the total cost of the five design alternatives included in Table 6.2.

Case' set of model parameters (i.e., 7.8 and 5.1 expected days of non-compliance in a year corresponding to Alt4 and Alt5, respectively).

6.3.5 Total cost estimates

The CapdetWorks (CapdetWorks, 2018; Harris *et al.*, 1982) software was used for calculation of the total cost corresponding to the different design alternatives. The calculated costs are based on the costing database for 2013 in the United States with 8% interest rate and 40 years for the lifetime of the project. They include operational, maintenance, materials and energy and capital costs. Figure 6.10 illustrates the relationship between the PONC values and the corresponding total cost for the different design alternatives.

Plotting the variation of PONC against the total cost can help designers identify those regions in design space for which the ratio of reduction in PONC to the increase in the total cost is at its highest and the effluent standards are met with a tolerable PONC. For example, if designers were interested in a NH_4 PONC value of less than 5% and a total cost in the range of 6 million dollars for the 'Worst Case' set of model parameters, Alt4 would be selected as the best design alternative.

6.4 SUMMARY

The application of the methods illustrated in this chapter provides additional insights into traditional approaches. The advantage of the proposed methods can be summarised as follows:

(1) They reduce subjectivity in the selection of design values, especially in situations where the engineers do not have enough experience (e.g., not enough knowledge on the effect of different process configurations on treatment performance).
(2) They provide an explicit, quantitative measure of compliance to effluent standards.
(3) They assist design engineers in identifying the limits of a specific treatment technology or process configuration as well as the design regions where the increase in certain process unit size would not result in a significant increase in the probability of compliance.

REFERENCES

ATV. (2000). Standard ATV-DVWK-A 131E, Dimensioning of Single-Stage Activated Sludge Plants. German Association for Water, Wastewater and Waste, Hennef, Germany.

Belia E., Benedetti L., Cierkens K., De Baets B., Flameling T., Nopens I. and Weijers S. (2012). Translating safety factors to sources of variability and uncertainty; Model-based design evaluation for the Eindhoven WWTP. Proceedings 85th Annual WEF Technical Exhibition and Conference (WEFTEC2012), September 29–October 3, 2012, New Orleans, Louisiana, USA.

Brun R., Kuehni M., Siegrist H., Gujer W. and Reichert P. (2002). Practical identifiability of ASM2d parameters – systematic selection and tuning of parameter subsets. *Water Research*, **36**(16), 4113–4127.

Bürger R., Diehl S. and Nopens I. (2011) A consistent modelling methodology for secondary settling tanks in wastewater treatment, *Water Research*, **45**(6), 2247–2260.

CapdetWorks v 4.0. (2018). State-of-the-art software for the design and cost estimation of wastewater treatment plants – user's manual. Hydromantis Environmental Software Solutions, Canada.

Cierkens K., Plano S., Benedetti L., Weijers S., de Jonge J. and Nopens I. (2012). Impact of influent data frequency and model structure on the quality of WWTP model calibration and uncertainty. *Water Science and Technology*, **65** (2), 233–242.

Harris R. W., Cullinane M. J., Jr. and Sun P. T. (1982). Process Design and Cost Estimating Algorithms for the Computer Assisted Procedures for Design and Evaluation of Wastewater Treatment Systems (CAPDET),

Environmental Engineering Division Environmental Laboratory, U.S. Army Engineer Waterways Experiment Station, Vicksburg, Mississippi.

Henze M., Gujer W., Mino T., Matsuo T., Wentzel M. C., Marais G. v R. and Van Loosdrecht M. C. M. (1999) Activated sludge model No.2D, ASM2D, *Water Science, Technology*, **39**(1), 165–182.

Hauduc H., Rieger L., Ohtsuki T., Shaw A., Takács I., Winkler S., Héduit A., Vanrolleghem P. A. and Gillot S. (2011). Activated sludge modelling: development and potential use of a practical applications database. *Water Science and Technology*, **63**(10), 2164–2182.

Menniti A., Schauer P., Naef R., Cullen N., Johnson B. and Chicoine L. (2014). Using Probabilistic Methods to Estimate the Risk of Nitrifier Washout and Set the Design SRT. Proceedings 87th Annual WEF Technical Exhibition and Conference (WEFTEC2014), September 27–October 1, 2014, New Orleans, Louisiana, USA.

Scheible O., Mulbarger M., Sutton P., Simpkin T., Daigger G., Heidman J., Yoder M., Schwinn D. and Storrier D. (1993) Process design manual: Nitrogen control. EPA/625/R-93/010 (NTIS PB94159142).

Schilperoort R. P. S. (2011). Monitoring as a Tool for the Assessment of Wastewater Quality Dynamics. Water Management Academic Press, Delft, The Netherlands.

Sin G., Gernaey K. V., Neumann M. B., van Loosdrecht M. C. M. and Gujer W. (2009). Uncertainty analysis in WWTP model applications: a critical discussion using an example from design, *Water Research*, **43**(11), 2894–2906.

Talebizadeh M. (2015). Probabilistic Design of Wastewater Treatment Plants. PhD thesis, Département de génie civil et de génie des eaux, Université Laval, Québec, QC, Canada.

Talebizadeh M., Belia E. and Vanrolleghem P. A. (2016). Influent generator for probabilistic modelling of nutrient removal wastewater treatment plants. *Environmental Modelling & Software*, **77**, 32–49.

Chapter 7

The bigger picture

7.1 INTRODUCTION

The previous chapters discussed how to use models for the evaluation of risk in engineering projects. They covered the identification and classification of the sources of uncertainty, their prioritisation and quantification and the methods by which we can incorporate them into a modelling project. However, the execution of engineering projects entails additional sources of uncertainty and risk.

This chapter provides a wider perspective of risks that can impact important decisions in infrastructure projects. Uncertainty and risk in Water Resource Recovery Facilities (WRRF) design can be analysed not just through the lens of a modelling project, but also through the lens of the project phase, stakeholder involvement or project delivery method:

- Project phase (Section 7.2): The degrees of freedom change dramatically depending on the project phase. Especially at the early stages of a project many decisions need to be made that have a huge impact on the final outcome. Uncertainty associated with these early decisions remains an issue that to date has received little attention.
- Stakeholders (Section 7.3): When planning, designing or operating treatment plants, various stakeholders become involved in the decision-making process. The involvement of these stakeholders may occur at different times during project development. Each stakeholder may bring a unique perspective of project uncertainty and will bring his/her own attitude towards delineating between acceptable and unacceptable risks. This creates uncertainty in how these conflicting perspectives are eventually resolved.
- Project delivery method (Section 7.4): Project delivery methods can distribute risk in different ways amongst stakeholders. In a design–build–operate (DBO) situation, all risk is borne by a single entity. In contrast, one party may be given a contract to design the project, a second party the contract to build, and a third party a contract to operate. Decisions will be influenced by each party's natural incentive to maximise project benefits for itself and to minimise the project downsides.

Editor(s): Evangelia Belia, Lorenzo Benedetti, Bruce Johnson, Sudhir Murthy, Marc Neumann, Peter Vanrolleghem and Stefan Weijers
doi: 10.2166/9781780401034_0111

These three ways of framing and the way they are interrelated are examined in more detail in the following sections.

7.2 ENGINEERING PROJECT PHASES

7.2.1 Overview

Figure 7.1 captures how decisions taken at different project phases – from the regulatory to the construction phase – impact the final process design (configuration and sizing) of a plant. This impact is proportional to the amount of uncertainty involved in the decisions taken in each of these phases. The graph also shows that following start-up, during commissioning and operation, additional decisions need to be taken that impact efficiency in plant operations.

In the permit specification phase (regulation), the regulator must define the-end-of-pipe requirements, thus determining plant effluent concentration and load limits. The effluent permit is a major driver in plant sizing and technology selection.

In the planning stage, the owner typically specifies the service-life time, the location, and the design flow and loads. Together with the regulatory requirements, the decisions in the planning stage are of major influence for the final design.

Given these criteria, engineering consortia will compete at the level of preliminary design or detailed design. It is at this stage that the process engineers are responsible for finding an optimal process solution given the regulatory permit and the boundary conditions specified in the planning stage. The degrees of freedom associated with uncertainty are the choice of a technology, the process configuration, parameter values for process models, among other things.

During the detailed design phase these choices are further refined, down to the detailed construction and implementation plans.

The construction phase typically does not have a large influence on decisions affecting the process design of the plant. During start-up and commissioning, the plant may not work as intended (e.g., non-ideal mixing or flow splitting), thus again increasing the degrees of freedom. Decisions need to be taken on how to adjust operations (e.g., operational set-points) to meet the intended plant performance.

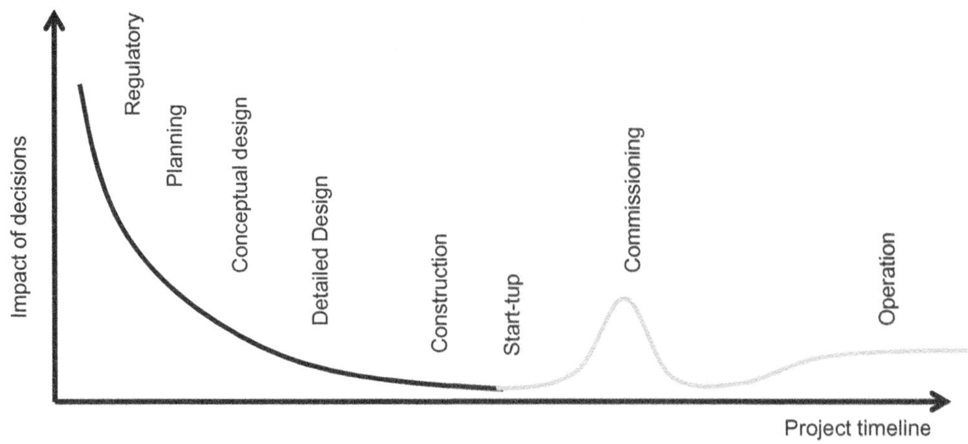

Figure 7.1 Impacts of decisions on plant process design are dependent on the project phase. Black line: decisions occurring from conception to start-up. Grey line: decisions occurring following start-up to continuous operation.

Table 7.1 Typical engineering project phases.

Project Phase	Definition
Regulatory phase	Defining treatment plant permits based on water quality considerations driven by local, regional and national legislation.
Planning	Developing the overall criteria for a facility, such as location, flows, loads, effluent quality, biosolids disposal, resource recovery and project time horizon. May include conceptual level unit process configuration. Conceptual level capital and operating costs are normally developed.
Preliminary design	Developing the overall concepts for a facility which includes control philosophy, process flow diagram, unit process sizing, and development of approach for support disciplines such as electrical, mechanical, structural, odour control, and site. Often considered approximately 10% of the total design effort.
Detailed design and construction	Producing the final design documents for all aspects of the facility, followed by the construction of the facility/improvements. Detailed design is sometimes split up into multiple phases, such as schematic design (30% of the total design effort), design development (60%), and construction documents (100% of the total design effort). Normally also includes start-up and troubleshooting of the new facilities.
Operations	The new facilities are operated by the permanent plant operations and maintenance staff to meet the regulatory requirements imposed on the facility

In the operations phase of a project, the risks are qualitatively different in that the owner is interested in minimising both the operating cost and as well as the risk of effluent non-compliance at the same time. The WRRF becomes an adaptive system and continuous changes to the initial design will take place during the infrastructure's lifetime. Table 7.1 includes more details on the tasks included during each project phase.

The degrees of freedom are the decision variables of each project phase. By making design decisions in the different project phases, the degrees of freedom are reduced throughout the project (Figure 7.2). If, in a

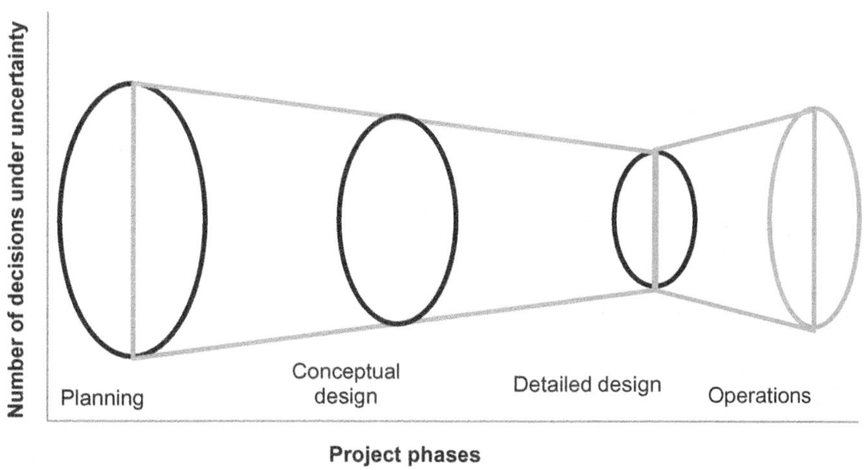

Figure 7.2 The number of decisions under uncertainty are reduced as the project progresses and increase again following plant start-up.

specific phase, a decision is made, that degree of freedom is eliminated and is considered as given in the subsequent project phases. The uncertainty associated with that degree of freedom is not always reduced or eliminated but is no longer considered. For example, design flows and loads are typically decided at the end of the planning phase. In the detailed design phase, the design flows and loads are typically assumed as given, even though uncertainty may still exist regarding their actual values. Table 7.2 includes examples of the typical degrees of freedom and the project phase where they are fixed.

Table 7.2 Typical degrees of freedom and phase where they are fixed (P = planning, PD = preliminary design, DD = detailed design, O = operation).

Phase/Degree of Freedom/Decision Variable	Project Phase where Degrees of Freedom are Fixed	Project Phase where Uncertainty is Evaluated
Plant location	P, PD	
Load and flows	P	PD
Temperature profiles		P, PD
Output requirements	P	PD
Definition of desired reliability/allowable risk	P, PD, DD	PD, DD, O
Technology pre-selection	PD	PD
Budget estimation	P	PD, DD
Technology selection	PD	
Process unit dimensions (preliminary)	PD	DD
Aeration capacity	PD	DD
Operational targets	PD, DD, O	
Chemical selection and dosing	PD, DD	DD, O
Number of reactors	PD	DD
Process unit dimensions (as builds)	PD	DD
Number and capacities of pumps	PD	DD
Number and capacities of blowers	PD	DD
Aeration system, number of diffusers	PD	DD
Redundancies	PD	DD
Mechanical equipment and redundancy	DD	
Electrical equipment and redundancy (UPS)	DD	
Control system and instrumentation	PD	DD
Operation of process equipment	PD	O
Set-points of automatic control loops	PD, DD	O
Software	PD, DD	O

7.2.2 Regulatory phase

Effluent criteria, which are key drivers for both the design and operation of treatment plants, are established by regulators. These criteria are either technology-driven or water-quality driven (e.g., Lijklema *et al.*, 1993). The criteria typically target the minimisation of acute toxicity, chronic toxicity or nutrient loading. Using water quality modelling and dilution calculations, site-specific WRRF permits are obtained.

Deriving these permits involves decisions which are subject to (sometimes significant) uncertainty. Various complex decisions are required in determining appropriate permit requirements for the WRRF that will protect the beneficial uses designated for the receiving water body. Normally, the final permits are a combination of effluent concentrations and load limits, either averaged over variable time limits or using statistical approaches such as 95th percentiles and medians. Although the decision processes of regulators are not the central focus of this STR, it is important to acknowledge that safety considerations take place when developing permits. In some cases, it is impossible to reach water quality objectives. In these cases, technology-based effluent limits are set.

Being responsible for the effluent discharge limits, the regulator assumes the risk that the effluent limits will maintain or improve the quality of the receiving water body. The assumption in this case (and by extension the risk), is that the information upon which the limits are based on is correct.

7.2.3 Planning phase

In the initial planning phase, the owner (typically with the assistance of a consulting engineer) makes choices that will heavily determine the final design. These choices need to deal with uncertainty in future loading, design life, costs and expected performance.

During the planning effort, the uncertainty considered by the designer/engineer and the owner is primarily associated with the development of flow and loading projections for a given facility, as well as the future effluent requirements. Uncertainties in flow projections are due to changes in population, rainfall, and changes in inflow/infiltration in the collection system. Uncertainties associated with loading projections are linked to changes in industry and population behaviour. Water conservation programmes for example, can impact hydraulic loading projections due to flow reductions to the wastewater treatment plants.

Moreover, uncertainty may be associated with the future impact on receiving water quality or ecology; often, safety factors are (implicitly) introduced here. The degree of treatment required is also subject to change over the life cycle of a project, and various scenarios must be considered during the planning process.

The sources of uncertainty of most importance to the owner are budget availability, changes in city/county/state design standards, and environmental requirements related to current and future regulations.

During this stage, both owners and designer/engineers need to review and understand uncertainties to ensure that client goals and technical requirements have been met. Typically, this is done by evaluating a range of possible scenarios and developing a path forward that addresses the needs of the stakeholders.

At the end of the planning phase, the degrees of freedom are reduced due to the decisions made and as a result several sources of uncertainty will not be considered in the subsequent project phases.

7.2.4 Preliminary (conceptual) design

In the preliminary design stage, the process engineer proposes a technology, the layout and sizing of the plant, as well as design effluent target levels. Design guidelines, simulators and costing tools are applied by the engineer to find the best solution, given the requirements and constraints set during the planning phase.

The choice of technology will depend on the criteria, the estimated costs and the engineer's familiarity and confidence in the technology. In the case of bioreactor selection, a first decision could be whether to use conventional suspended activated sludge treatment, a membrane or a biofilm system. In the next step, a basic process configuration is decided upon. This includes such decisions as the use of step-feed, the number of treatment lines, the number of bioreactors and their basic geometry.

The sources of uncertainty during preliminary design are related to the assumptions needed to develop the conceptual approach and hydraulic profile. The engineer typically has to make assumptions about the chemical/physical properties of the wastewater and its effects on unit process performance. Examples include the influent profiles for wastewater flows and loads.

Tansel (1999) states that uncertainties are introduced into the design process as a result of gaps between available and needed information at different points of the design process. This often leads to plants that are 30–50% overdesigned based on municipal codes and, after safety factors are used by designer/engineers, are overdesigned by 100% or more (Russell, 2006).

Other sources of uncertainty for the designer/engineer are associated with the veracity of the existing as-built information (for upgrade projects), topographical mapping, and geotechnical report.

The goals of this phase are typically technology selection, plant dimensioning of bioreactors, clarifiers, and other unit processes as well as the evaluation of aeration capacity, and the determination of all major control loops. In addition, the associated instrumentation and chemicals to be used are selected.

7.2.5 Detailed design, construction, and start-up

The detailed design stage deals with issues related to equipment redundancy and the selection of mechanical and electrical equipment in view of robust and safe operations. Regarding uncertainty and risk, it is in this step that reliability engineering gains importance.

Where the goal of preliminary design is to refine the design criteria and concepts initially established in the scope of work, the purpose of detailed design is to produce the final contract drawings including the plans, specifications and any other supporting documents. The number of uncertainties which have not been addressed by the designer/engineer is reduced at this design stage with the goal of managing any remaining uncertainties during construction.

The goal of the detailed design phase is to develop the conceptual design to the level of detail required for plant construction. Fixed degrees of freedom at this stage are final reactor and other unit dimensions, volumes, required flows, aeration capacities, chemical dosage type and amounts.

During the last decade, a shift has occurred, from identifying the main source of uncertainty as the kinetic parameters, to influent variability and dynamics and proper model structure of transport physics, such as mixing, aeration and sedimentation as well as chemical processes like precipitation. Modelling scenarios focus on better aeration distribution modelling, improved clarifier modelling with computational fluid dynamics (CFD) and controller models (Nopens *et al.*, 2015; Rehman *et al.*, 2017; Sin *et al.*, 2008).

The main uncertainties that are dealt with in the detailed design phase relate to process reliability as a result of equipment reliability and redundancy.

The determination of the number, sizing and configuration of equipment is made in view of reliable plant operations. At this stage, robustness and redundancy are considered. Precautions are taken to provide adequate treatment under malfunction as well as maintenance scenarios. Malfunction scenarios include failure of equipment, such as pumps, valves, aeration equipment. A typical maintenance scenario is a tank being out of service due to cleaning or repair.

Blower configuration and tank geometry, inlet and outlet structures are specified to guarantee optimal transport and mass transfer. CFD, introduced above, is a methodology that can assist the engineer with this.

Also, the design of robust Instrumentation/Control and Automation (ICA) equipment is considered at this stage. Methods include failure detection and the use of soft sensors.

Although construction itself includes many risks, they are not strongly related to the final plant configuration. However, during start-up, new degrees of freedom may be introduced. It is here where some of the assumptions made in the design are tested and some decisions may need to be revised: for example, sludge settleability, inhibition of organisms, obtaining the required population of microorganisms in the bioreactors, non-ideal mixing or flow splitting, among others.

7.2.6 Operations

After the commissioning phase, a treatment plant is not typically running at the design load but will be initially under-loaded. From this point onwards, the plant becomes an adaptive system (Dominguez & Gujer, 2006; Neumann *et al.*, 2015). As part of a robust design, plant operating strategies will be preliminarily defined by the design engineer. Operations staff then determine how best to run the facility within the designed constraints. For example, a facility might be designed to operate as a plug flow system under normal conditions but switched to step feed for wet weather conditions. Daily decisions are made on how much sludge to waste, how much chemicals to dose, how to time digester supernatant return, and so on. The daily operation includes the management of problems such as bulking and foaming, toxic inflows and equipment failure. A typical longer-term decision the operator needs to take is how close to the permit to run the plant. This will typically depend on the penalty scheme in place related to permit compliance (e.g., incrementally increasing taxation, binary penalty (pass/fail) or penitentiary sentence). Incentives such as maintaining prestige may also be present. Figure 7.3 gives an

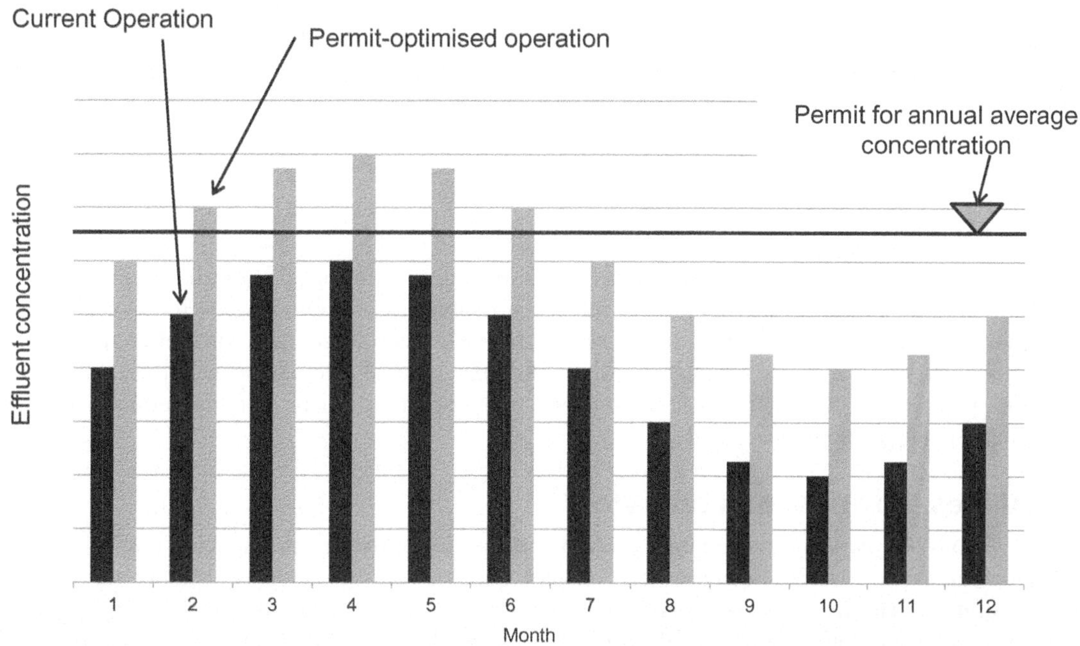

Figure 7.3 To safeguard against violating the annual effluent permit the plant is operated so that (black bars) the plant effluent concentration is permanently (e.g., for the maximum month) below the legal annual permit (bold horizontal line). The grey bars indicate how it may be possible to operate the plant, and still meet the annual permit.

example in which an operator runs the plant at a high margin of safety in the presence of a yearly average concentration limit. To safeguard against violating the annual effluent permit, the operator aims (black bars) to maintain the plant effluent concentration below the legal yearly permit in every month. In this way, the operator eliminates the risk of not complying. The grey bars indicate how it may be possible to run the plant, and still meet the yearly limit. However, this strategy implies that the operator could predict the loads and plant performance for the coming months with a small margin of error.

A feasible strategy will lie between the two extremes. It is one objective of the Task Group to highlight the importance of considering how the incentives of different stakeholders may lead to a risk-taking or risk-averse strategy.

Table 7.3 summarises the typical project phases, the uncertainties associated with each phase, the decisions that need to be taken under uncertainty, the expected deliverables of each project phase and examples of how models can be used to assist in uncertainty evaluations.

7.3 STAKEHOLDERS

7.3.1 Overview

When planning, designing or operating treatment plants, various stakeholders make decisions at different points in time. Uncertainty is associated with the degrees of freedom available when making those decisions. This is a function of how far project development has progressed (see Section 7.2). As the project moves forward, certain decisions either become irreversible (like the regulator has made the decision on permit level) or they can only be reversed at tremendous cost. The associated uncertainties can be removed from the analysis.

Project decisions vary with each stakeholder. The regulator needs to decide on the plant's permit. The planner needs to specify a design horizon and an associated design load. The design engineer needs to make assumptions on the current and future wastewater composition; he needs to choose an appropriate configuration as well as a process technology and values for the associated parameters. The operator needs to decide on how close to the permit to run the plant. Finally, the owner needs to decide on an upgrading and investment strategy. Depending on the contract type, the stakeholder sequence may differ.

7.3.2 Regulators

Translating water quality objectives into WRRF permits is associated with considerable uncertainty. It is not uncommon for safety factors to be included at this stage of the process. Often some form of pollution allocation takes place when the WRRFs in the same watershed need to comply with different permits. Exposing the rationale of these regulatory decisions may reveal alternative solutions for water infrastructure planning at the watershed level (e.g., set bubble permits where a single permit is set that covers multiple plants within a watershed).

7.3.3 Utilities – owners and operators

Utilities' decisions that pertain to uncertainty can range from the normal decision making that is a part of everyday operations to strategic management choices such as bid selection or finding an optimal investment strategy. The risks and benefits from these decisions accrue at the level of individuals. For example, if the plant operator lowers dissolved oxygen in the plant to reduce energy costs, she may not be acknowledged for the associated benefits even though she has increased her highly visible risk of not meeting effluent limits. Clarifying how incentives and penalty schemes affect individuals and their decisions is therefore a basis for modifying behaviour.

Table 7.3 Typical project phases, associated uncertainties and examples of how models can be used to assist in uncertainty evaluations.

Project Phase	Risk/Uncertainties	Decisions	Deliverables	Model Implementation Examples	
				Model Use	**Key Sources of Uncertainty**
Regulatory	Information upon which the limits are based on is correct	Permit limits	Permit limits	Simulate receiving water body quality	Flows and loads Measured data
Planning	**Designer:** Flow and load projections Future effluent requirements **Owner:** Budget availability Changes in design standards Environmental requirements related to current and future regulations	Future wastewater infrastructure Plant location Technology selection Wastewater flows/loads during dry/wet weather Performance requirements (extent of treatment)	Capital improvement plan Specifications Location selection Budget planning	Future scenario evaluation Technology investigation Checking if future output requirements are achievable	Flows and loads Boundary conditions (i.e., temperature profiles)
Preliminary design	**Designer:** Wastewater chemical/physical properties Influent flows/loads/characterisation Variability in influent flows / loads Data quality Veracity of the existing as-built information (for upgrade projects) Topographical mapping Geotechnical report	Design inputs Safety factors Process design parameters Selection of effluent design criteria Selection of design values for unit processes and mechanical equipment	Technology selection Process configuration Sizing Layout Capital costs Operational costs	Plant dimensioning Performance evaluation System selection System optimisation Control system design Selection of sensors, actuators and locations	Flows and loads Model structure Model parameters Influent fractionation Mass transfer model Kinetic Stoichiometric Actual (imperfect) flow distribution
Detailed design, construction, start-up and commissioning	Project management Low bid environment and poorly written specifications Lack of flexibility in design features Errors and omissions in contract documents Cost estimating errors by designer/engineer Equipment reliability/redundancy Process reliability from equipment reliability and redundancy	Schedule Selection of mechanical and electrical equipment Quality control of documents and plant systems installed Design change orders during construction Manual control handles Automatic control loops Fall-back procedures	Final contract drawings Plans, specifications and other supporting documents Design of instrumentation/control and automation (ICA)	Impacts of tanks out of service Impacts of equipment malfunction CFD modelling	Aeration system design Required equipment redundancy to achieve requested reliability CFD models to design flow patterns and proper mixing
Operation	**Designer**: Proper implementation of any control systems **Owner:** Compliance Financial risk of power/chemical use, mechanical failures and maintenance	Operational decisions	Operational risk management action plans	Process optimisation Controller settings Limits of performance Debottlenecking Operational strategies Performance benchmarking Post project audits Impact of failures on effluent quality Redundancy evaluations	Equipment failures Unforeseen weather Toxic spills Pandemics

7.3.4 Engineers

Inherent variability in the inputs to a WRRF, along with uncertainty in the value of parameters critical to design, complicates the engineer's effort to design a system that will produce an effluent of acceptable quality at all times. The incentive to minimise the probability of a performance failure creates incentives for oversizing the system. WRRFs will always be subjected to unanticipated events that are difficult to design for.

In-depth knowledge about uncertainty and variability and how to successfully address them can give an engineering company a competitive advantage and help owners better understand the proposed designs. As the degrees of freedom increase when moving from guidelines to mechanistic model-based design, addressing the uncertainties and the associated risks becomes more important for engineers.

7.3.5 Public

A common assumption is that communicating the risk involved in engineering projects will reduce public trust. However, a lack of systematic research makes it difficult to evaluate such claims. Van der Bles *et al.* (2020) found that transparency on issues of uncertainty does not harm the public's trust in the facts or in the source. On the contrary, people 'can handle the truth' about the level of certainty or uncertainty in scientific facts and knowledge. Based on their results, the authors recommend that the communication of uncertainty in the media is best conveyed through numerical ranges with a central point estimate. This format, in particular, did not seem to significantly influence (i.e., reduce) perceived trust and reliability in either the number or the source of uncertainty. In addition, they draw attention to the fact that using the word 'estimate' or increasing the magnitude of the confidence interval did not seem to alter people's perception of uncertainty, which points to the need to better contextualise the degree of uncertainty for people.

A key challenge to maintaining public trust in science is for communicators to be honest and transparent about the limitations of the current state of knowledge.

7.4 CONTRACT DELIVERY METHODS

7.4.1 Overview

In infrastructure procurement, the contract delivery mechanism determines how risk is allocated among different stakeholders. Depending on the way the infrastructure procurement contract is set-up, risk will be allocated differently to the owner, engineer or contractor. The type of contract determines who is going to profit from the opportunities and who is going to bear the cost of possible failures. For instance, a consortium competing in a design bid might want to optimise between not being sued due to proposing a plant that turns out to be under-sized and not losing the bid to a competitor due to being too conservative in the choice of the values for the design inputs. It is obvious that the incentives for different stakeholders are dependent on how risks and opportunities are shared in these contracts. Molenaar *et al.* (2004) discuss the risk allocation among stakeholders for different contract types in the wastewater sector.

7.4.2 Examples of delivery methods

A common method of project delivery method is the design–bid–build (DBB). In a DBB delivery, the owner normally hires a designer/engineer to develop project documents. Once the design is complete, the owner bids the work and hires a contractor to construct the project. The successful bidder then builds the project, with oversight by the owner and (normally) the designer/engineer. In this delivery method the owner and the contractor each assume cost risks. Most of the cost risk is assumed by the owner and builder. The designer/engineer assumes only a small amount of the cost risk.

For DBB-type projects, the owner, and to some degree the engineer, take responsibility for the influent parameter selection. In this case it is the engineer's responsibility to provide the owner with adequate information to make informed decisions about the design parameters, and their impacts upon the project.

Alternate delivery methods, such as design–build (DB), and design–build–operate (DBO) have become increasingly popular to owners because these delivery methods shift in varying degrees the financial and process risk to the contractor. In addition, they also move the risk to the party that is best able to balance the various process-related and financial risks. The alternate delivery methods often result in a different analysis of risks than a conventional design approach.

Additional discussion on these project delivery methods has been included in Appendix E.

7.4.3 Stakeholder involvement as a function of contract type

Table 7.4 illustrates the involvement of different stakeholders as a function of contract type and project phase.

In many parts of the world, traditional procurement in the water sector has relied on DBB contracts. In this example, based on water quality considerations, a regulator (R) will develop an effluent permit for the new treatment plant. The utility (U) together with the municipality (M) and a private company (P0) will develop the requirements for the future plant, possibly in collaboration with the regulator.

Usually, the utility will hire a private company P1 to assist in developing the preliminary and detailed design. This is followed by a tender for the construction of the plant which is then carried out by construction company P2. The commissioning of the new plant will typically be undertaken by both the design company P1 and the construction company P2. Then, the utility (U) will operate the plant.

During the past 30 years, there has been a steady increase in public–private–partnerships (PPP) leading to many different types of delivery mechanism. This has led to the rise of DB contract where consortia bid for both the design and the construction. In some cases, it may also include operation (DBO) and in some other cases, also ownership is transferred for a pre-determined length of time (design–build–own–operate–transfer: DBOOT). Some of these different schemes are discussed in

Table 7.4 Stakeholders responsible for taking decisions within the project phases for two contract delivery methods.

	Delivery Method	
Project Phase	**DBB**	**DBO**
Regulatory	R	R
Planning	P0, U, M, R	P0, M, R
Preliminary design	P1, U	**P1**
Detailed design	P1, U	**P1**
Construction	**P2**	**P1**
Commissioning	P1/P2	**P1**
Operation	U	**P1**

P0: private company, U: utility; M: municipality; R: regulator. The indices 1 and 2 in P differentiate between different companies. In bold the phases covered by the actual contract.

more detail in Appendix E. Appendix E also includes examples of the types of project delivery methods used in several countries across the world.

7.5 SUMMARY

Engineering decisions taken under uncertainty are heavily influenced by the contractual environment, the role of the stakeholders and the phase of an infrastructure project. This chapter discussed how these play a far greater role in shaping the final outcome of an infrastructure project than is widely acknowledged.

REFERENCES

Dominguez D. and Gujer W. (2006). Evolution of a wastewater treatment plant challenges traditional design concepts. *Water Research*, **40**(7), 1389–1396.

Lijklema L., Tyson J. M. and Lesouef A. (1993). Interactions between sewers, treatment plants and receiving waters in urban areas: a summary of the Interurba '92 workshop conclusions. *Water Science & Technology*, **27**(12), 1–29.

Molenaar K. R., Bogus S. M. and Priestley J. M. (2004). Design/build for water/wastewater facilities: state of the industry survey and three case studies. *Journal of Management in Engineering*, **20**(1), 16–24.

Neumann M. B., Rieckermann J., Hug T. and Gujer W. (2015). Adaptation in hindsight: dynamics and drivers shaping urban wastewater systems. *Journal of Environmental Management*, **151**, 404–415.

Nopens N., Laurent J., Wicklein E., Ducoste J., Griborio A., Batstone D., Wicks J., Saunders S., Potier O., Ratkovich N., Portela A. M. and Randal W. S. (2015). Computational Fluid Dynamics: an important modelling tool for the water sector. New Developments in IT & Water, 2nd Conference Rotterdam, The Netherlands.

Rehman U., Audenaert W., Amerlinck Y., Maere T., Arnaldos M. and Nopens I. (2017). How well-mixed is well mixed? Hydrodynamic-biokinetic model integration in an aerated tank of a full-scale water resource recovery facility. *Water Science & Technology*, **76**(8), 1950–1965.

Russell D. (2006) Practical Wastewater Treatment. Wiley and Sons Publishing, Hoboken, NJ, USA.

Sin G., De Pauw D. J. W., Weijers S. and Vanrolleghem P. A. (2008). An efficient approach to automate the manual trial and error calibration of activated sludge models. *Biotechnology and Bioengineering*, **100**(3), 516–528.

Tansel B. (1999). Propagation of parameter uncertainties to system dependability. *Civil Engineering and Environmental Systems*, **16**(1), 19–35.

van der Bles A. M., van der Linden S., Freeman A. L. J. and Spiegelhalter D. J. (2020) The effects of communicating uncertainty on public trust in facts and numbers. *Proceedings of the National Academy of Sciences of the USA (PNAS)*, **117**(14), 7672–7683. https://doi.org/10.1073/pnas.1913678117 (accessed 2021).

Chapter 8

Perspectives

8.1 INTRODUCTION

This chapter seeks to provide a vision on how uncertainty and variability may be handled in the future. We structure the main discussion along the major project phases: regulation, planning, design and operation. A central objective followed by the Task Group has been very technical in nature: 'Replacing the safety factor-based approaches with methods that account for uncertainty in explicit ways'. While investigating the feasibility of such approaches, the Task Group has encountered broader implications. These implications extend far beyond the scope originally set out by the Task Group, which was essentially limited to the explicit inclusion of uncertainty in models describing the physical and bio-chemical processes occurring in treatment plants. In this chapter, a vision integrating the above-mentioned broader implications is developed, by formulating two general objectives and then, listing methods that support these within the different project stages.

The two general objectives that underlie the quest for uncertainty-based methods are:

- **Explicitness**: In current approaches, variability and uncertainty are mostly handled in an implicit way, that is, they are lumped into a safety factor approach. In the future, the Task Group envisions that the sources of uncertainty and variability are made explicit and when possible quantified.
- **Transparency**: In current projects, the rationale of how a design was selected – a process that involves taking decisions under uncertainty – is not always available. A more detailed documentation of decisions in design and optimisation projects as well as in bid selection would improve transparency. This could also be supported by post-project audits to measure the long-term success of a design. A continuous monitoring of loads, capacity and performance of a plant and comparison to planning and design assumptions may also lead to the adoption of alternative service delivery mechanisms or alternative infrastructure configuration and reduce planning biases of future projects.

Editor(s): Evangelia Belia, Lorenzo Benedetti, Bruce Johnson, Sudhir Murthy, Marc Neumann, Peter Vanrolleghem and Stefan Weijers
doi: 10.2166/9781780401034_0123

8.2 SOCIOECONOMICS AND APPLIED MATHEMATICS

The Task Group believes that a combination of methods from both socioeconomics as well as applied mathematics and statistics will be helpful in realising the two objectives presented in the introduction.

8.2.1 Socioeconomics

The wastewater engineering field has successfully integrated the natural sciences (physics, chemistry, biology). Can we assume a similar integration with socioeconomic sciences in the near future? New approaches such as fore-sighting or multi-attribute decision theory may enhance the toolbox of the engineer and lead to new ways of thinking about infrastructure planning and design. Specific methods that might be used include:

- **Fore-sighting methods**: The adoption of exploration tools such as structured scenario analysis could be especially useful in the early planning phases of wastewater treatment plant projects (Schoemaker, 1995). Spending more time and resources in developing a wide range of potential storylines on how the catchment could develop into the future can improve the credibility of the entire project (e.g., Dominguez *et al.*, 2011) and create better support from all stakeholders.
- **Life-cycle assessment (LCA)**: Including sustainability criteria, for example, through an LCA means that projects can be judged more holistically than in current practice which often focuses on cost–performance criteria (e.g., Corominas *et al.*, 2013; Renou *et al.*, 2008), leading to a more transparent and explicit trade-off between different objectives.
- **Multi-attribute-utility theory**: Such tools (Clemen, 1996) structure decision-making processes in order to reach 'optimal' decisions under multiple criteria (e.g., Reichert *et al.*, 2007) enabling a smoother decision-making process.
- **Benefit–cost–risk approaches**: Making explicit the incentives of different stakeholders and showing how benefits, costs and risks are spread among them can improve overall infrastructure provision.
- **Environmental economics**: Valuing the benefits of preventing pollution, and maintaining ecosystem services go beyond the fulfillment of discharge limits and can make the services provided by wastewater treatment more visible to society.
- **Benchmarking and auditing**: Independent reviews of a city's infrastructure performance and the comparison of results to other cities can identify best practice approaches in view of handling or reducing uncertainties.

8.2.2 Applied mathematics and statistics

For the more detailed technical design stages as well as for the optimisation of operations, a more rigorous use of applied mathematics and statistics offers significant potential for managing both uncertainty and variability in more explicit ways. The increase in computer efficiency results in faster processing of larger problems. This means that one can compute more complex models (computational fluid dynamics, plant-wide models, integrated models including sewer systems and rivers), perform long-term dynamic simulations, apply more sophisticated techniques from systems analysis and artificial intelligence (probabilistic procedures, data mining) or introduce real-time systems for model-based predictive control. All of these advances address various sources of uncertainty and variability. When applying these methods, the wastewater engineer/modeller will need to make a trade-off between rigour and pragmatism in deciding which uncertainties are relevant and need to be considered explicitly in a quantitative fashion, as considering all possible uncertainties is not feasible nor desirable.

8.3 ACCOUNTING FOR UNCERTAINTY IN PROJECTS

The following sections assess how methods from both socioeconomics and applied mathematics can be applied to different stages of wastewater infrastructure development projects.

8.3.1 Regulatory phase

At several instances, the Task Group was confronted with the question: 'What about uncertainty/variability concerning WRRF effluent requirements?' Two main issues are of interest. The first relates to the uncertainty in the derivation of effluent requirements while the second concerns how different effluent permits treat uncertainty and variability when assessing compliance.

Considerable uncertainty is involved in establishing water quality targets to protect a receiving water and as a result their determination often evokes debate. Tools from environmental and ecological economics may offer support to regulators, increase transparency on the benefits of a specific target and facilitate comparisons of infrastructure cost vs. environmental benefits.

Once water quality targets have been determined, they need to be translated into effluent permits. With regard to these effluent permits, the Task Group found considerable differences in their formulation, especially between North America and Europe. In North America, often criteria such as average yearly concentration or maximum monthly concentration are prevalent. When performing model-based design, it is not evident how to define the critical scenarios.

In Europe, percentile-based requirements are used mostly (e.g., 90th percentile day, 50th percentile day). These types of permits account for temporal variability in a straightforward way when using dynamic simulation for design. This permit formulation also facilitates explicit uncertainty evaluation and thus enables more realistic project design.

8.3.2 Planning phase

A successful development of the urban wastewater infrastructure requires continuous interaction with urban planning departments.

Uncertainty is reduced through good relationships between different departments, such as between the infrastructure- and urban planning departments. Then issues can be addressed in a way that leads to integration on questions such as: 'what type of urban development facilitates sustainable wastewater management?' or 'what type of wastewater infrastructure best serves the intended urban development?'

As the life span of a WWTP is in the order of 25 years, uncertainties in the planning phase can be considerable. Scenario analysis is the most widespread tool to deal with these issues (e.g., Dominguez *et al.*, 2009).

The introduction of new infrastructure procurement methods such as service contracts can also be used to change the agent responsible for decisions on planning variables that contain uncertainty. Making explicit how benefits, cost and risk are spread among different stakeholders could improve the formulation of tender requirements and thereby attempt to quantify the uncertainty involved (Flyvbjerg *et al.*, 2003).

8.3.3 Preliminary design

Multi-attribute-utility methods, LCAs and benefit–cost–risk analyses can be ways to make uncertainties explicit and visible at this stage. For the dimensioning of reactors, the Task Group envisages an increased use of complex models for plant-wide design as well as the application of probabilistic procedures.

An example of a desired output of a probabilistic procedure is given in Figure 8.1. The example shows how an engineering consultant may obtain a least-cost design within a 'Design–Build–Operate' bid. The

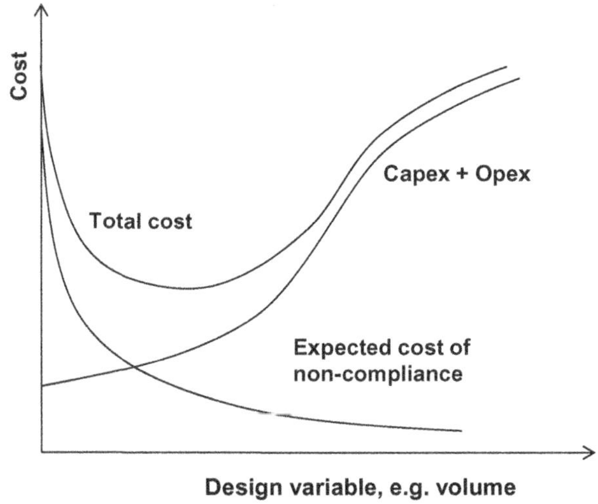

Figure 8.1 Optimisation exercise of an engineering company in a Design-Build-Operate bid. The total cost is the sum of capital and operational expenditure (Capex + Opex) and 'expected' cost of non-compliance (= cost of non-compliance × probability of non-compliance).

company tends to consider capital- and operational expenditure (Capex + Opex) which is with increasing safety (e.g., larger tank volume). Being able to quantify the probability of non-compliance as a function of a design variable enables the estimation of the expected cost of non-compliance as a function of that design variable. In this way, a design with lowest total cost can be identified.

8.3.4 Detailed design

Since unit operations are interconnected, procedures are needed to analyse how the design of a particular unit operation might impact the design of a different unit operation and the overall system performance. The design engineer would benefit from having a tool built upon a knowledge base that would automate a sensitivity analysis that would consider how the individual pieces affect the performance of a whole. The sensitivity analysis would lead to an initial version of the final design that would then require further refinement.

The use of computational fluid dynamics may offer significant potential for optimisation of the flow and transport characteristics of reactors (such as preventing dead-zones and hydraulic short-circuits).

In this design stage, reliability engineering is expected to gain importance in the future for the structural, mechanical and electrical engineering domains to guarantee redundancy in view of robust operations (Sharma *et al.*, 1993; Tung *et al.*, 2006). The design of an Instrumentation Control and Automation (ICA) system can be supported using reliability modelling of actuators (pumps, compressors, diffusers) and sensors.

8.3.5 Operation

The Task Group suggests the introduction of continuous post design-project audits in order to continuously reduce uncertainty. This requires the use of robust sensors for continuous performance assessment and the reporting of non-compliance events in order to correlate failures with their causes: for example, is a non-compliance event due to system size, due to the malfunctioning of a sensor or a controller, due to the inhibition of microorganisms from toxic discharges or the unintended functioning of a process (e.g., foaming or bulking)?

Modern data mining techniques can support the understanding of the plant reliability and the identification of failure events (Duerrenmatt & Gujer, 2012). Process auditing could be done at different time scales: high-frequency data (e.g., 15-minute intervals) to yearly performance measures. Monitoring at high frequency can identify critical reliability issues and improve operational strategies, such as ICA and facilitate the use of model-predictive control. It may also yield insights into the impact of other factors, like human/operator behaviour on plant performance.

The analysis at larger timescales (Dominguez & Gujer, 2006) can uncover issues related to planning assumptions: for example how realistic were the wastewater load projections? Did innovations in technology allow a more efficient use of the installed infrastructure? Did the introduction of new permits render the initial design obsolete? Often, the initial assumptions and the decision process leading to a design are not revisited once the plant is in operation. Performance audits performed across many plants would identify strengths and weaknesses of different procurement strategies and through feeding benchmarking studies help to identify best practices for WWTP planning and design. Such long-term studies could also point to the potential of alternative infrastructure settings and strategies discussed in the following section.

8.4 ALTERNATIVE WAYS OF HANDLING UNCERTAINTY

One of the central objectives of this STR is to propose methods that expose and where possible quantify sources of uncertainty to increase transparency. An alternative to trying to better account and quantify uncertainty is to construct systems that are less vulnerable to the actual sources of uncertainty and variability. These include systems that change when conditions change (adaptive systems) or that are capable of absorbing alternative outcomes (robust systems).

Adaptivity of systems in view of uncertainty can be attained by increasing flexibility, modularity or decentralisation. To increase flexibility, methods (such as 'real options') have been suggested to quantify the value of higher up-front investments that significantly decrease costs for possible expansions that might become necessary in the future (Gersonius *et al.*, 2013). Increasing modularity is another option, where a treatment system is made up of single modules with shorter service lifetimes that can be exchanged more easily. Finally, decentralisation of treatment systems offers another pathway to increase adaptability. For instance, in the case of a one-plant-per-building approach, the uncertainty in the planning stage (such as population growth/shrinkage) and its relevance to the required capacity would be almost completely eliminated. However, the uncertainty linked to performance and the resulting receiving water quality may increase if there is a lack of professional supervision typically present at larger scale plants. An additional disadvantage of small decentralised plants is that, adapting one large plant may be easier than adapting very many small ones, especially when trying to improve effluent water quality, decreasing GHG emissions or implementing resource recovery.

Robustness means that systems are conceived that are not looking to be optimal for an expected outcome, but that work satisfactorily for many possible conditions or future. Robust systems include systems that can switch between different regimes. Such systems would typically require higher implementation of instrumentation, control and automation.

8.5 OUTLOOK

In the ideal approach, most uncertainties are discoverable and can be expressed in some type of mathematical/statistical formulation that can be considered during model simulation. The information needed to achieve this can be extensive. In the end, the practice of probabilistic designs will at best be an approximation of this ideal. In addition, not all uncertainty can be expressed in a statistical sense.

Fore-sighting, such as scenario analysis will necessarily be part of the process. Even so, there will always remain recognised ignorance and total ignorance that by their nature live outside the realm of uncertainty and scenario analysis and can only be observed with hindsight, that is, after having occurred. Nonetheless, by combining the lessons learnt from past plant design and operational experiences with the knowledge and tools that are currently available, significant improvements can be achieved.

REFERENCES

Clemen R. T. (1996). Making Hard Decisions: An Introduction to Decision Analysis. Duxbury Press, USA, August 15, 1997; ISBN-13: 978-0534260347.

Corominas L., Foley J., Guest J. S., Hospido A., Larsen H. F., Morera S. and Shaw A. (2013). Life cycle assessment applied to wastewater treatment: state of the art. *Water Research*, **47**(15), 5480–5492.

Dominguez D. and Gujer W. (2006). Evolution of a wastewater treatment plant challenges traditional design concepts. *Water Research*, **40**(7), 1389–1396.

Dominguez D., Worch H., Markard J., Truffer B. and Gujer W. (2009). Closing the capability gap: strategic planning for the infrastructure sector. *California Management Review*, **51**(2), 30–50.

Dominguez D., Truffer B. and Gujer W. (2011). Tackling uncertainties in infrastructure sectors through strategic planning: the contribution of discursive approaches in the urban water sector. *Water Policy*, **13**(3), 299–316.

Duerrenmatt D. J. and Gujer W. (2012). Data-driven modelling approaches to support wastewater treatment plant operation. *Environmental Modelling & Software*, **30**, 47–56.

Flyvbjerg B., Bruzelius N. and Rothengatter W. (2003). Megaprojects and Risk – an Anatomy of Ambition. Cambridge University Press, Cambridge, UK.

Gersonius B., Ashley R., Pathirana A. and Zevenbergen C. (2013). Climate change uncertainty: building flexibility into water and flood risk infrastructure. *Climatic Change*, **116**(2), 411–423.

Reichert P., Borsuk M., Hostmann M., Schweizer S., Spoerri C., Tockner K. and Truffer B. (2007). Concepts of decision support for river rehabilitation. *Environmental Modelling & Software*, **22**(2), 188–201.

Renou S., Thomas J. S., Aoustin E. and Pons M. N. (2008). Influence of impact assessment methods in wastewater treatment LCA. *Journal of Cleaner Production*, **16**(10), 1098–1105.

Schoemaker P. (1995). Scenario planning – a tool for strategic thinking. *MIT Sloan Management Review*, **36**(2), 25–40.

Sharma M., Hall W. and McBean E. (1993). Reliability analyses of activated sludge systems in attaining effluent standards. *Canadian Journal of Civil Engineering*, **20**(2), 171–179.

Tung Y. K., Yen B. C. and Melching C. S. (2006). Hydrosystems Engineering Reliability Assessment and Risk Analysis. McGraw-Hill, New York, USA, p. 512.

Appendix A

Terms and definitions – application and discussion

A.1 INTRODUCTION

Discussions on the topic of uncertainty can be confusing as similar terms are often used interchangeably for different concepts. This Appendix defines and discusses the concepts and terms related to uncertainty within the context of model-based design and operation of wastewater treatment systems. This Appendix is organised in four sections.

The first section includes terms commonly used in modelling.

The second section covers general terms relating to basic statistical concepts and metrics that form the basis for all uncertainty evaluations. The section lists a set of general terms that describe relationships between measured and simulated quantities. Knowledge of these terms is necessary as they form the basis of all the key concepts in the field of uncertainty. For each term, a general definition has been included and how the term is applied to measurements, model structure and parameter values and results of model simulation and prediction.

The third section covers the most essential concepts and terms regarding uncertainty.

The final (fourth) section presents comparative discussions on terms and concepts confounded with uncertainty.

Editor(s): Evangelia Belia, Lorenzo Benedetti, Bruce Johnson, Sudhir Murthy, Marc Neumann, Peter Vanrolleghem and Stefan Weijers
doi: 10.2166/9781780401034_0129

A.2 MODELLING

Table A1 Terms and definitions relating to model architecture.

Area	Term	Definition	Example
Model architecture	Model	Abstract mathematical description of a system.	The ASM1 model.
	Variable (of system)	Changing characteristic of a system.	A biomass or substrate concentration
	State (of model)	Model states describe system variables.	Bulk liquid ammonia concentration, symbolised as SNH.
	Dynamic model	Model describing the evolution in time of variables of interest ($dx/dt = f(x, t)$).	The ASM1 model describes temporal changes in variables such as COD, oxygen and ammonia (among others).
	Steady-state model	Model describing the expected values of variables of interest under fixed conditions for process inputs and operational variables ($dx/dt = 0$).	Total COD = C1 + C2 + C3, pH = log(H +)
	Algebraic state	State which is computed from other states by means of an algebraic (non-dynamic) equation	Acid–base equilibrium equations in some applications
	Parameter	Value that specifies the behaviour of a system or system model. A parameter is considered to have a single true value only for a given system and/or model, although it may be unknown a priori, and can be determined separately for each application of a given model structure.	Maximum growth rate, affinity constants, kLa
	Stoichiometric parameter	Numerical relationship between compounds participating in a given (bio)chemical reaction or the conversion rates of a compound relative to those of another compound.	Yield
	Kinetic parameter	A parameter included in the equation describing the rate by which a given (bio)chemical reaction occurs as a function of one or more model states	Maximum growth rate, affinity constant
	Constant	Value that specifies the behaviour of systems or models in a universal and invariant way. A constant remains unchanged throughout all applications of a model	Gravitational constant, molar masses

Table A2 Terms and definitions relating to acts of modelling.

Area	Term	Definition	Example
Acts of modelling	Simulation	Generation of a model response	Effluent ammonia concentrations are simulated
	Forecasting	Generation of a model response for future conditions based on currently available information	Expected future nitrogen removal rates are evaluated by means of prediction of nitrogen concentrations in the effluent
	Model set-up or structure identification	Definition or adjustment of the model structure including: reactor design, reactor configurations, reactions and biokinetic model structure	Fixed-film or suspended growth reactor, anoxic and aerobic zones, ASM-type or model using other growth/inhibition terms.
	Calibration or parameter optimisation	Adjustment of parameters for a given model structure so to better reflect observations made in a specific set of conditions (operation)	Observations made during a measurement campaign are used to fit the model to the data
	Validation	Comparison of the simulation results of a calibrated model and an independent set of observations. In theory, the data used for validation contains no information contained in the data used for calibration.	Industrial practice: validation is executed for a wide range of purposes. Good statistical practice: validation is exclusively aimed at the selection of a better/best model structure.
	Objective (function)	Degree of performance of a given model. An optimal model is a model with the best performance. The same objective is used throughout calibration, validation and testing steps.	Often a least-squares objective is used (e.g., mean-squared residual, MSR), in which the objective (MSR) needs to be minimised for model improvement.
	Identifiability	*Ability to assign a unique, optimal value to the model parameters* under reasonable expectations for data availability and data quality. A model lacks structural identifiability when one cannot assign unique values to the model parameters even in the hypothetical case with an infinite amount of data that is representative (e.g., including dynamics) and is perfectly accurate (no bias, no variability). Parametric identifiability concerns the idea that the parameters of a structurally identifiable model are not necessarily identifiable in practically realisable situations with finite resources and practical limits on available dynamics and data quality.	
	Domain of validity or generalisation	A set or range of situations, either foreseen or not, under which a given model is (still) applicable. A good quality of the model is the ability to stretch its use or extrapolate it further compared to other models. The domain of validity can also be seen as the set of conditions within which a model will give results reliable enough to serve as a basis for a decision, despite its uncertainty. This ability can be defined as generalisation properties or domain of validity. The term domain of validity suggests that a distinct quantifiable boundary for generalisation exists.	
	Optimality	An optimal model is a model that is the best among those available in reaching a certain objective. Such objectives may range from describing a certain phenomenon qualitatively to predicting concentrations of interest.	

Table A3 Terms and definitions relating to model evaluation.

Area	Term	Definition	Example
Model evaluation	Least-squares (LS)	Least-squares objectives/optimality refer to the practice of using a penalty function which is a sum of squared prediction errors/residuals used to calibrate, validate and select models. The use of a LS objective is usually motivated based on the assumption that measurement errors are independently and identically distributed according to the normal distribution.	Sum of squared residuals (SSR), mean -squared residual (MSR) and root mean-squared residual (RMSR).
	Sum of squared residuals (SSR)	Sum of squared residuals.	$SSR = \sum_{i=1}^{m} (r_i)^2$ for $i = 1 \ldots m$ residuals
	Mean-squared residual (MSR)	Average of squared residuals.	$MSR = \frac{1}{m} \times SSR$ for $i = 1 \ldots m$ residuals
	Root mean-squared residual (RMSR)	Average of squared residuals.	$RMSR = \left(\frac{1}{m} \times SSR\right)^{1/2}$ for $i = 1 \ldots m$ residuals
	Sum of absolute residuals (SAR)	Sum of absolute residuals.	$SAR = \sum_{i=1}^{m} \lvert r_i \rvert$ for $i = 1 \ldots m$ residuals
	Mean absolute residual (MAR)	Average of absolute residuals.	$MAR = \frac{1}{m} \times SAR$ for $i = 1 \ldots m$ residuals
	Independently and identically distributed (i.i.d.)	Notion that a set of given outcomes (e.g., prediction residuals) are distributed independently and characterised by the same distribution. Independence practically means that the value of one outcome is not informative about another outcome. I.i.d. conditions are typically assumed for measurement errors.	
	Normal distribution	The normal or Gaussian distribution is a widely assumed and applied distribution for residuals and errors and can be characterised by two parameters, namely mean and standard deviation. The probability density function follows a symmetric bell-shape.	

A.3 STATISTICS

Table A4 Terms and definitions relating to basic statistical concepts and metrics.

Term	General Definition	Measurement	Model Structure/Parameters	Model Simulation/Prediction
Outcome	Result of measurement, experimentation, simulation or modelling.	For example, a nitrate concentration.	–	For example, an OUR estimate, a biomass concentration prediction, a kinetic parameter estimate.
Measurement	Assessment of the value of a variable of interest by means of an analytic experiment or on-line signal generation.	For example, a dissolved oxygen measurement.	–	–
Error	Deviation between an outcome and its true value.	Numerical difference between a measurement and the true corresponding value in the sampled system.	Difference between the true system and the model representation. This can be in structure and parameters (separately or simultaneously).	Difference between a predicted value and the reference value in the modelled system or a reference value.
Residual	Deviation between an outcome and its reference value.	–	–	Difference between a predicted and a measured concentration.
Credibility	Probability or degree of belief that a given outcome corresponds to its true, usually unknown, value.	Probability or degree of belief that a given measurement reflects the true underlying variable well.	Probability or degree of belief that a model is representative for the true system.	Probability or degree of belief that a simulated result corresponds well to the true corresponding value.
Credible interval/region	Interval within which an outcome or the region within which a set of simultaneous outcomes are believed to lie, with a given probability.	Range around a measurement within which the true value is expected to lie with a given level of confidence.	Range of model structures and/or parameters within which the true system is expected to be with a given level of confidence.	Range around a simulated result within which the true corresponding value is expected to lie with a given level of confidence.
Confidence interval/region	Interval in which an outcome or the region in which a set of simultaneous outcomes are to be found with a given frequency, when repeated many times	Range within which a repeated measurement is expected to lie with a given level of probability	Range of model structures and/or parameters that will be reached with a given frequency upon repetition of the applied data collection, model identification, and/or calibration procedures	Range of model structures and/or parameters that will be reached with a given frequency upon repetition of the applied data collection, model identification, calibration, and/or simulation procedures

(Continued)

Table A4 Terms and definitions relating to basic statistical concepts and metrics (*Continued*).

Term	General Definition	Measurement	Model Structure/Parameters	Model Simulation/Prediction
Bias or systematic error	Bias is the consistent deviation of the measured value from an accepted reference value (ISO 15839:2003). In statistical texts, bias and systematic error are considered to be one and the same.	Bias is introduced into measured variables by means of consistent error (s). It is recommended to use measurement bias to explicitly refer to bias in a measurement device or outcome.	Bias in model structure selection or model parameter identification is the systematic deviation between the real system and the model representation. This concurs when the considered model(s) are not representative of the system or when the data used for model identification, calibration, and validation is biased.	Bias in model simulation is the systematic over- or under-prediction of a variable of interest as the model is unable to sufficiently predict the observations made.
Trueness	Antithesis of bias, that is, the degree of how close an outcome is to an accepted reference value.	How close the measurement is to its reference value.	–	–
Precision	Precision is the closeness of agreement between independent measured values obtained under stipulated conditions (ISO 15839:2003). Precision is a qualitative concept and not a number.	–	Degree to which repeated modelling exercises will deliver a similar model.	Closeness of independently reproduced outcomes under the same, specified conditions to each other.
Variability	Spread of 'true' values of a quantity. In measurements, variability is the opposite of precision (ISO 5725-1:1994). Variability is an expression of random error and is a property of the population, not of our state of knowledge (Kelly and Campbell, 2000).	Degree to which repeated measurements show different or dissimilar results; also, the degree of being far from each other.	–	–
Accuracy	Comprises trueness and precision and is therefore a single expression for systematic and random error.	Closeness of agreement between a measured value and the accepted reference value (ISO 5725-1:1994, ISO 15839:2003).	–	–

A.4 UNCERTAINTY

Table A5 Terms and definitions relating to essential concepts regarding uncertainty.

Term		Definition
Uncertainty		Degree of inability to determine or predict the exact behaviour of a system or process both now and in the future. Uncertainty relates to (1) the inability to determine truly and precisely what has happened in the past because several possibilities lead to similar observations and (2) the inability to predict truly and precisely what will happen in the future. Uncertainty results from lack of knowledge and is *partly* reducible through the acquisition of additional knowledge, for example, more data or further understanding of a process.
Nature of uncertainty	Aleatory (irreducible) uncertainty	Aleatoric uncertainty is representative of unknowns that differ each time the same experiment is run. It is due to the inherent variability of a system and cannot be reduced with any further research (e.g., rainfall, toxic spills). It is classified as irreducible and called variability.
	Epistemic (reducible) uncertainty	Epistemic uncertainty is due to things that could be known in principle but are not known in practice. This may be because a quantity has not been measured sufficiently accurately, or because the model neglects certain effects. It can be reduced with further research or measurements (e.g., experimental determination of kinetic parameters) in which case it is classified as reducible and called epistemic uncertainty.
Level of uncertainty	Quantifiable uncertainty	Can be quantified and described with statistical methods and can be attributed to uncertainties such as a random measurement error of a sensor.
	Scenario uncertainty	Can be described with qualitative estimations of possible outcomes that may develop in the future. Realistic assumptions about relationships and/or driving forces within the model can be established. It is not possible, however, to derive the probabilities of the scenarios taking place.
	Recognised ignorance	State where fundamental uncertainty exists, and the scientific basis is insufficient to develop functional relationships, statistics or scenarios.
	Total ignorance	State where the actors are not aware of uncertainty. It is unknown what is unknown.
Location (source) of uncertainty	Context uncertainty	Context refers to the economic, political, social and technical conditions and circumstances that influence the model boundaries and frame the issues that the model is to address. Context uncertainty also relates to the suitability of a model for its intended purpose.
	Input uncertainty	Includes system data uncertainty and external driving force uncertainty. Data uncertainty includes uncertainty in, for example, the influent flow and concentrations to a model. External driving force uncertainty relates to uncertainty associated with changes in conditions that are outside the model boundaries but rather are inputs describing the reference system and external forces driving changes in the current system.
	Model uncertainty	Both model structure uncertainty and model numerical uncertainty arising from computer implementation of the model.
	Parameter uncertainty	Parameter uncertainty is associated with the lack of knowledge regarding the true value of the model parameters as well as uncertainty associated with parameter optimisation technique used during model calibration (e.g., lack of convergence, parameter selection for optimisation).
	Model output uncertainty	The total uncertainty assessed by uncertainty propagation taking all model uncertainties into account.

A.5 DISCUSSION OF TERMS OFTEN CONFOUNDED WITH UNCERTAINTY

A.5.1 Precision and variability

Precision is the quality of a repeated process or procedure to deliver similar results. Variability is the degree of absence of that quality. The larger the variability, the lesser the precision. The term variability is recommended to describe the concept qualitatively while standard deviation and variance are measures used to quantify variability (Taylor & Kuyatt, 1994).

A measurement is variable when subjected to random disturbances or fluctuations. An example that is easily demonstrated is that of a noisy sensor. Even in lab conditions one expects different values for repeated measurements. A less obvious example is sampling error which can induce variability. Indeed, one does not always sample the exact same volume of water or at the exact same spot. Heterogeneity of the medium may cause variation as well.

Precision of simulation results is the degree to which several simulation results are similar to each other. For model quantities (e.g., influent) that are variable, a simulation result can be generated for each possible value of that quantity. A distribution for the simulated variable can be generated based on a single simulation only, usually based on the mean parameter value and transformed into a confidence interval for interpretation. As such, the confidence interval quantifies the precision/variability quality of the simulation.

A.5.1.1 Quantification of precision and variability

Precision should not be defined as the inverse of standard deviation (Taylor & Kuyatt, 1994). Precision is therefore a qualitative concept and not a number. It is most typical to quantify variability (imprecision) rather than precision (ISO 15839:2003). To this end, it is common to estimate the standard deviation. Root mean square residual (RMSR) is the most popular approach to estimate the standard deviation. For this, one subtracts reference values and/or (estimated) bias from the measurements and then computes the averaged square of these residuals. This is then a measure of spread of the measurements.

For simulation results, variability is obtained in a similar way as for measurements. One subtracts reference values and (estimated) bias and then measures the spread, for example by computing RMSR.

A.5.2 Accuracy and uncertainty

An outcome that is close to its reference/true value is more accurate; one that is further away is inaccurate or uncertain. Accuracy comprises trueness and precision and is therefore a single expression for systematic and random error. For this reason, accuracy should only be used as a qualitative concept (Taylor & Kuyatt, 1994) and one should avoid quantifying it. Instead, accuracy/uncertainty should be described with separate measures of bias and variability.

Improving both trueness and precision simultaneously to any desired degree is generally impossible, thereby resulting in a necessary compromise. Since it is not defined how this compromise should be made a priori and since measures for trueness and precision are typically in different scales, it is most often left to the end-user to make this trade-off.

The accuracy of a measurement is the closeness of the given value to the true value. Measurement uncertainty is thus the degree of inability of the measurement to describe the true corresponding value.

The accuracy of a model is the closeness of the model to the described true system. Naturally, model uncertainty is then the inability of the model to describe the targeted system well. Model accuracy/uncertainty can be quantified in several ways depending on the information available. In the purest sense, it is quantified based on the mismatch between the model (structure and parameters) and the true system from which data was derived. More practical measures are based on the ability of the model to predict the true behaviour of the system. A model validation or test step (see below) can serve this purpose.

A.5.2.1 Quantification of accuracy and uncertainty

Accuracy and uncertainty are difficult to quantify for several reasons. Firstly, because the true values are not available for real systems. This can be accommodated in practice by using reference values (see Section A.5.6). Secondly, because accuracy/uncertainty encompasses both systematic as well as random deviations from the truth. Quantifying both by means of one single measure is difficult and has little value in view of model improvement as systematic and random deviation requires different actions for model improvement. As such, accuracy should be decomposed into trueness and precision when attempting quantification. Similarly, uncertainty should be decomposed into bias and variability for purposes of quantification.

A.5.3 Error and residual

The term error is recommended to describe the difference between the obtained value and the corresponding true value and residual for the difference between the obtained value and the reference value if this reference is different than the true value. Except for well-designed laboratory experiments, only residuals are available in practice. Figure A.1 illustrates this notion where the obtained value is given by model simulation (ySIM) and the reference value is a measurement (yREF) for a true value (yTRUE).

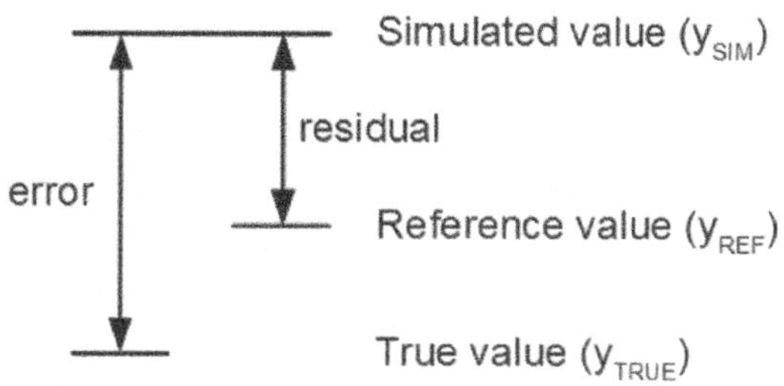

Figure A.1 Suggested separation of true value and reference value and, correspondingly, error and residual.

A.5.4 Trueness and bias

An outcome that is generally or systematically close to its reference/true value is truer; one that is generally or systematically further away is biased. In this definition, the terms generally or systematically are important as they define the difference with accuracy/uncertainty. Conceptually, trueness/bias describes the consistent, general, or long-term characteristics of an outcome.

Upon repetition (of a measurement or experiment), an averaged outcome will (typically) converge to a value called the expected value. The difference between this expected value and the reference value is an estimate of the bias. As such, bias expresses systematic error. The larger the bias, the smaller the trueness. Bias is introduced into measured variables by means of consistent error(s). In Taylor and Kuyatt (1994), bias is reserved exclusively to measurements and systematic error is considered a generally applicable term. [ISO 5725-1:1994; ISO 15839:2003] do not make such a distinction. This may be due to sampling at a location that systematically results in values too high or too low for the overall system. Other possibilities include sensor calibration errors, systematic error in the analytic protocol, and systematic error in handling of samples and/or results.

Bias in model structure selection or model parameter identification results when the considered model(s) are not representative for the system. In other words, insufficient flexibility of the model(s) to be fit to a given data set results in a systematic deviation between the real system and the model representation. For example, kinetic reaction rate coefficients may be far from reality, a one-step nitrification model may be consistently off if the second step nitrification is relatively slow. Another reason for model bias is that the data used for model identification, calibration, or validation was biased.

Bias in model simulation is recognised as the systematic over- or under-prediction of a variable of interest as the model is unable to sufficiently predict the observations made. This may be due to errors introduced throughout measurement campaigns or in modelling or, alternatively, introduced in the measurement of the variable of interest. It may also be due to the extrapolation of a model to a situation which it was not calibrated for or it could be that the system behaviour has changed since the measurement campaign used for modelling.

A.5.4.1 Quantification of trueness and bias

It is most common to quantify bias, rather than trueness. Bias can be quantified relatively easily for measurements. For this, one repeats the measurement (same sampling and measurement procedure) several times and compares to a reference value. The distance between reference and the average of the measurements is an (least-squares optimal) estimate of the bias. The reference value may consist of a trusted reference measurement (e.g., standard protocol) or the set concentration in lab-made standards or samples.

For model structure/parameters, it is difficult to estimate bias, as it requires a true or reference model, which generally is not available. As a result, bias in the model is usually ignored in practice, implicitly assuming that model structure and used data are unbiased. One way of assessing model bias, is to compare simulation results to real measurements to evaluate whether consistent deviations are present. The validation or testing step (see below) may serve that purpose.

For simulation results, one compares simulation results with reference values and computes the consistent deviation between the two. A practical measure is the mean deviation between simulation result and reference value.

A.5.5 Note on true values

Many of the definitions presented in Appendix A include the term 'true value'. True values are generally not known and by virtue of this, definitions based on the knowledge of 'true' values are of little practical use. It is therefore common to replace 'true values' with 'accepted reference values' (ISO 5725-1:1994; ISO 15839:2003; Taylor and Kuyatt, 1994). It is however crucial to realise that one then deliberately ignores the mismatch between accepted reference values and true values as a source of uncertainty. Whether this is of importance will depend on the quality of the accepted reference values. According to [ISO 5725-1:1994; ISO 15839:2003], an accepted reference value is:

(a) An assigned or certified value based on experimental work of some national or international organisation;
(b) A consensus or certified value based on collaborative experimental work;
(c) A theoretical or established value based on scientific principles;
(d) When (a), (b) and (c) are not available, the expectation of the (measurable) quantity, that is, the mean of a number of measurements.

The use of accepted reference values is common for sensor calibration. For example, one obtains laboratory standard measurements and uses these to check a sensor which is based on a different measurement principle. This is unlikely to be useful for model parameters. For predictions, one may be able to obtain standardised measurements as a reference. Importantly, when one uses reference values, one inherently assumes that these show no bias or variability of their own, that is, one considers the reference values perfect.

A.5.6 Note on repetitions

Variability of measurements, parameter estimates and model predictions can be described by means of repetitions. It is assumed that these repeated values are produced in the exact same way each time (identically distributed) and that they are independent of each other (independent sampling). In practice, this may not be the case for the following reasons:

- There is simply no repetition made;
- The measured/parameter/prediction value depends on other values, for example, through redundant relationships or dynamic relationships. For example, it is common that two parameters are correlated. Also, consecutive measurements of a variable in a dynamic system are likely auto-correlated.

Without repetitions, one can do little to obtain separate descriptors of bias and variability. In the rare case where the true value is known, one can only obtain an overall measure of uncertainty based on a single measurement. In such a case, one has little clue on how one can reduce this uncertainty since the reduction of bias and variability require different actions.

One can try to explicitly account for dependencies between distinct variables through redundancy or dynamic relationships when one has measurement or values for these variables. For example, data reconciliation techniques may be used. To obtain corrected values which satisfy the assumed relationships, for example, a mass balance over a process unit, one uses the corrected values as reference values. Following that, one can compute residuals between measurements and corrected values, which then serve to characterise the residuals in terms of bias and variance. Here, one will now typically assume that the residuals are independent and identically distributed. Available techniques belong to the fields of statistical process control and data reconciliation. Note that one typically assumes that the relationships are given without error, that is, they are a perfect representation of reality.

A.5.7 Bias, variability and uncertainty: a graphical example

Consider that one aims at a target in a shooting game and one has multiple chances to try. At each trial, one may or may not be close to the target. After a series of trials, one can characterise the distribution of the result in the different trials. Suppose four people are participating in the game. For each individual, one obtains results as in Figure A.2. It is typical to describe these results in terms of bias and variance. The combination of both represents the uncertainty.

In the result on the top right of the figure, one has the results for the best player. This player has a low bias, that is, the average of all trials is close to the target. This player has also low variability since all trials are close to each other. The second player (top left) also has low bias (average of trials is on target) but shows much more variability.

At the bottom-right, the third player's results are shown. In this case, the average of the results is far from the off target. One says that the player shows bias. However, the trials are very close to each other, meaning that the player shows low variability. Finally, the average of results for the fourth player is off target while the trials are rather far from each other. This player thus shows high bias and high variability.

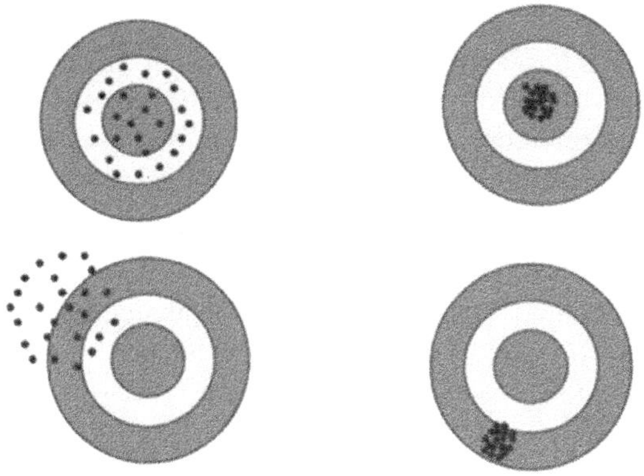

Figure A.2 Top left: low bias, high variability. Top right: low bias, low variability. Bottom-left: high bias, high variability. Bottom-right: high bias, low variability. (Source: http://www.minedesignwiki.org/index.php/Sampling_QAQC).

Uncertainty is now a combination of both bias and variability. As such, one can say that the first player shows low uncertainty. One also says that this person is accurate. The fourth player shows high uncertainty, or in other words, low accuracy. The second and third players show uncertainty levels between the first and fourth players' uncertainty. Importantly, it is not always clear how one should weigh accuracy against precision. It is difficult to gauge whether player 2 is better or worse than player 3.

A.5.8 Link between measurement, modelling and prediction

One can also characterise measurements, model (parameters) and prediction in terms of uncertainty, bias and variability. In these cases, a bias is the general tendency of a measurement, model parameter estimates or prediction to have a different value than the true value. Variability is the descriptor for how far repeated values for measurements, model parameter estimates or predictions are from each other. In Figure A.3, each square represents the average of three measurements demonstrating measurement error, the vertical dotted line represents the influent variability of the daily average, the horizontal dotted line shows the annual average and the continuous line shows the interannual variability of the measurement.

A.5.9 Qualitative model performance criteria

A.5.9.1 Identifiability

In certain situations, it may not be possible to obtain (good) estimates of parameters. This is usually the result of the combination of (1) a model with a rather large number of parameters which one aims to identify, (2) lack of representative, dynamic data or (3) lack of data quality. According to Dochain *et al.* (1995), identifiability can be defined as follows:

Assume that a certain number of the state variables are available for measurement; on the basis of model structure (structural identifiability) or on the type and quality of available data (practical identifiability), can we expect to give via parameter estimation a unique value to the model parameters?

A model lacks structural identifiability when one cannot assign unique values to the model parameters even in the hypothetical case with an infinite amount of perfect data that is representative (for example including dynamics) and is perfectly accurate (no bias, no variability).). In this case, additional data collection cannot aid in the modelling process. In contrast, parametric identifiability concerns the idea

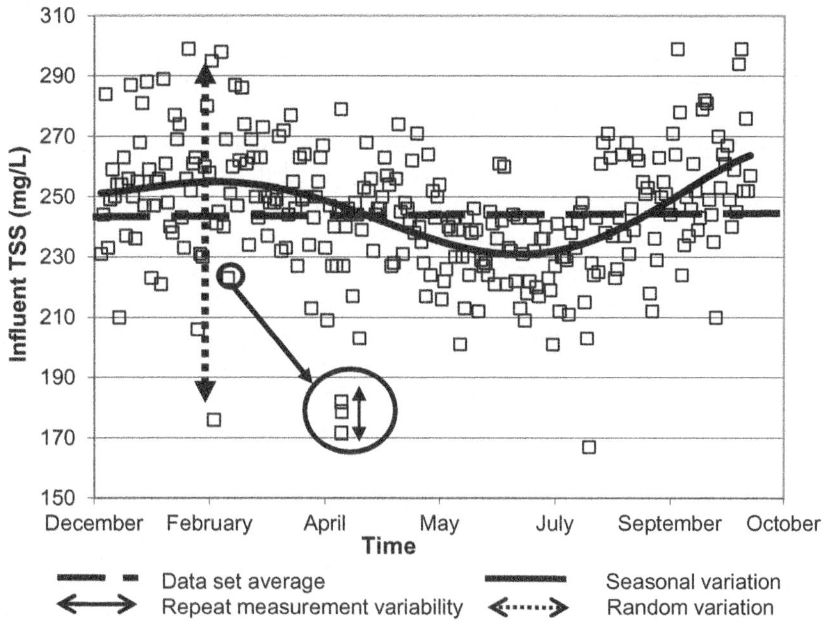

Figure A.3 Measurement error, daily average variability and inter-annual variability of influent TSS.

that the parameters of a structurally identifiable model are not necessarily identifiable in practically realisable situations with finite resources and practical limits on data quality.

A.5.9.2 Generalisation and domain of validity

As one may expect that during modelling, one has no access to data representative of all behaviours of the system, it is important to consider the capacity to extrapolate the model beyond the conditions covered in the calibration and validation set also, one may expect that the true system itself may change over time. Consequently, one will likely arrive in a situation where the model is used still while not exactly representative of the true system. Such extrapolation may be valid or not. A good quality of the model is thus the ability to stretch its use or extrapolate it further compared to other models. The domain of validity can also be seen as the set of conditions within which a model will give results reliable enough to serve as a basis for a decision, despite its uncertainties. This ability can be defined as generalisation properties or domain of validity. The latter suggests that a certain and quantifiable boundary for generalisation exists.

A.5.9.3 Optimality

An optimal model is a model that is the best among those available in reaching a certain objective. Such objectives may range from describing a certain phenomenon qualitatively to predicting concentrations of interest.

A.5.10 Reliability and redundancy

According to IEEE (1990), reliability is the ability of a system or component to perform its required functions under stated conditions for a specified period of time. For example, reliability of a model is the degree to which one is certain that a given model will support its intended task over a time horizon, that is, the extent to which one can rely on the model in the future. As such, reliability includes the notion that the performance of a model will degrade over time due to the inability to incorporate unforeseen changes in the represented systems or inherent incompleteness.

The basic method to assess reliability is to define the risk, define the likelihood, and define the consequence (preferably as a cost). This method has been used to assess criticality, which is used for capital improvements planning and prioritisation. The criticality of a project is the product of the likelihood/frequency of failure and the consequence of failure.

Redundancy in an engineered system can be defined as a design practice to include backup components in a system or incorporate interchangeable components so that the system can be repaired quickly. In either case, the intention is that the system can operate at an acceptable performance level without interruption when a piece of equipment fails or must be taken out of service. This is important in safety-critical systems (e.g., plane, nuclear reactor) to avoid damage in the case of failures in a single part of the system. Data redundancy is the degree to which multiple measurements contain the same information. This property is what one uses to remove gross outliers from data sets by means of mass balances. Most statistical techniques for fault detection are based on redundancy (e.g., Frank, 1990).

Wastewater treatment redundancy should not be confused with process reliability. Reliability refers to the inherent dependability of a piece of equipment, a unit process, or the overall treatment process in meeting the design objective (Tanaka *et al.*, 1998). In terms of effluent quality, Niku *et al.* (1979) and Bott and Parker (2011) referred to process reliability as the ability to meet the specified effluent requirements free from failure, or as the probability of success, where failure is the probability that the effluent concentration is greater than the discharge permit limit. McBride and Ellis (2001) defined reliability as the percentage of time a wastewater treatment plant remained in compliance with discharge standards. Reliability analysis has been used to predict the performance of a technology over time and to determine the strategies that improve performance and reduce risks of failure (Etnier *et al.*, 2005). Redundancy on the other hand, can be viewed as a subject of reliability. Redundancy is practiced in the design of wastewater treatment plants to improve reliability through the provision of standby equipment or processes that reduce the risk of failure to meet water quality regulations or guidelines (Palmer *et al.*, 2003).

Although the definition of redundancy does not include regulatory compliance or reliability standards, some data show a direct relationship between treatment process reliability, redundancy design and regulatory compliance. Bott and Parker (2011) concluded in a comprehensive study of nutrient removal plants, that one of the main causes affecting the performance of treatment plants was the reliability and redundancy of important unit processes or pieces of equipment in the wastewater treatment plants.

A.5.11 Robustness and resiliency

In general, one desires that any engineered system can handle disturbances for quite some time before (large) deviations in operation are seen. Robust systems do not easily break down or disintegrate, that is, they can withstand extreme conditions without visible changes in structure or functionality. This capacity is usually described as robustness which can be defined as the property that permits a system to maintain its functions against internal and external perturbations (Kitano, 2004). Robustness is a characteristic which becomes apparent when imposing extreme or potentially harmful conditions (stress) onto a good-working system. For example, concrete is a robust material as it does not change shape under considerable stress. Rubber is not a robust material as it changes form and shape under slight stresses already. Naturally, a larger robustness is generally a good thing. However, robustness generally comes at a cost. For example, in order to make a buffer tank more robust to extreme flows, one needs to design a larger tank.

Resiliency is a term used broadly and differently in different contexts. In the field of process control, it signifies a property very similar to robustness, for example, resilient control systems are those that tolerate fluctuations via their structure, design parameters, control structure and control parameters (Mitchell & Mannan, 2006). More recent definitions provided in the context of cyber-security, include a non-random component to the cause of disturbances and a goal of awareness: A resilient control system is one that maintains state awareness and an accepted level of operational normalcy in response to disturbances, including threats of an unexpected and malicious nature (Rieger *et al.*, 2009). Thus, to achieve resiliency, it is not sufficient to tolerate or cope with a disturbance, one must also be aware of it. This makes a clear distinction with robustness, which is a passive approach to handle (random) disturbances.

A general definition of resiliency could be the degree to and/or rate at which a system can recover from disturbances or upsets, caused by random causes or wilful actions. In contrast to robustness, which is characterised under stress conditions, resilience is, in addition, characterised by means of a recovery process or period. For example, can one get the initial performance again? Does the system return quickly to normal behaviour? Concrete can be considered a non-resilient material. Indeed, once broken it is not easy to fix and one will generally replace it with new concrete. Rubber is a resilient material. Indeed, as one releases stress, rubber generally goes back to its original shape and form. Resilience can also be regarded as a capacity of self-healing. As with materials, it is expected that robustness and resilience are to be bargained against each other. In addition, resilience also comes at a certain cost. For a buffer tank to recover faster from an extreme flow event, one may need pipes with a larger diameter.

REFERENCES

Bott C. B. and Parker D. S. (2011). Nutrient Management Volume II: Removal Technology Performance & Reliability, WERF Research Project under Nutrient Challenge. WERF Report NUTR1R06k.

Dochain D., Vanrolleghem P. A. and Van Daele M. (1995). Structural identifiability of biokinetic models of activated sludge respiration. *Water Research*, **29**(11), 2571–2578.

Etnier C., Willetts J., Mitchell C., Fane S. and Johnstone D. (2005). Decentralised Wastewater System Reliability Analysis Handbook. Project No. WU-HT-03–57. Prepared for the National Decentralised Water Resources Capacity Development Project, Washington University, St. Louis, MO, by Stone Environmental, Inc., Montpelier, VT.

Frank P. M. (1990). Fault diagnosis in dynamic systems using analytical and knowledge-based redundancy: A survey and some new results. *Automatica*, **26**(3), 459–474.

IEEE. Institute of Electrical and Electronics Engineers. (1990). IEEE Standard Computer Dictionary: A Compilation of IEEE Standard Computer Glossaries. New York, NY ISBN 1559370793

ISO 5725-1. (1994). Accuracy (trueness and precision) of measurement methods and results – Part 1: General principles and definitions. International Organisation for Standardisation, ISO 5725-1:1994.

ISO 15839. (2003). Water quality – On-line sensors/analysing equipment for water – Specifications and performance tests. International Organisation for Standardisation, ISO 15839:2003.

Kelly E. J. and Campbell K. (2000). Separating variability and uncertainty in environmental risk assessment – making choices. *Human and Ecological Risk Assessment: An International Journal*, **6**(1), 1–13.

Kitano H. (2004). Biological robustness. *Nature Reviews Genetics*, **5**(11), 826–837.

McBride G. B. and Ellis J. C. (2001). Confidence of compliance: a Bayesian approach for percentile standards, *Water Research*, **35**(5), 1117–1124.

Mitchell S. M. and Mannan M. S. (2006). Designing resilient engineered systems. *Chemical Engineering Progress*, **102**, 39–45.

Niku S., Schroeder E. and Samaniego F. (1979). Performance of activated sludge processes and reliability-based design. *Journal (Water Pollution Control Federation)*, **51**(12), 2841–2857.

Palmer T. M., Ross M. C. and Jank B. (2003). Efficient Redundancy Design Practices. International Water Association (IWA), London, UK.

Rieger C., Gertman D. and McQueen M. (2009). Resilient control systems: Next generation design research. 2nd Conference on Human System Interactions, pp. 632–636.

Tanaka H., Asano T., Schroeder E. D. and Tchobanoglous G. (1998). Estimating the safety of wastewater reclamation and reuse using enteric virus monitoring data. *Water Environment Research*, **70**(1), 39–51.

Taylor B. N. and Kuyatt C. E. (1994). Guidelines for Evaluating and Expressing the Uncertainty of NIST Measurement Results. National Institute of Standards and Technology Technical Note 1297, MD, USA. https://www.nist.gov/pml/nist-technical-note-1297 (accessed 2021).

Appendix B

Methods for uncertainty analysis

B.1 UNCERTAINTY FRAMEWORKS

B.1.1 Frequentist

In the frequentist framework, it is assumed that the true value of a quantity (e.g., model parameter) is fixed (Johnson & Wichern, 2007). The problem is then to find a good approximation of that value as well as a region of confidence where this true value lies. This confidence interval is a measure of the uncertainty about the true value of the quantity. The distribution describes the probability for the obtained value of a variable or parameter (e.g., of variable measurements, model parameter estimates).

In general, the frequentist framework advocates the repetition of experiments in order to obtain samples of measurements or parameter estimates. Based on this sampling, one then estimates the distribution of the obtained values for the variable or parameter. For example, one can calculate the mean and variance which offer a complete characterisation of the normal distribution, assuming that the choice of a normal distribution is correct.

In the application of frequentist theory, it is assumed that by increasing the size of the sample, the estimated distribution will converge to the true distribution. For this to be true, two important conditions need to be met. First, the sampling procedure must lead to independently sampled values for the quantified variables. This is not always the case, especially when dealing with dynamic processes. Often, one has no access to repeated measurements or parameter estimates which are independent from each other. Second, the true distribution function should be able to be described by means of the fitted distribution function. This is often violated as well. Indeed, it is typical to assume the normal distribution for parameters in models that are non-linear in the parameters while the true distribution is not normal.

B.1.2 Bayesian

In the Bayesian framework, it is assumed that the quantity one seeks to identify is uncertain. Therefore, the quantified variable can take several plausible values. Each of these values will appear with different

Editor(s): Evangelia Belia, Lorenzo Benedetti, Bruce Johnson, Sudhir Murthy, Marc Neumann, Peter Vanrolleghem and Stefan Weijers
doi: 10.2166/9781780401034_0145

probability. This probability can again be described by a distribution; however, this distribution describes the probability for *the true value of a variable or parameter* (e.g., of the *true* variable value, *true* model parameters).

In the Bayesian context, one assumes a process which generates data. This process is given as a model which includes parameters and possibly input values. With such a process model, one generally describes the likelihood of measurements, y, conditional to the model parameters, p, denoted $L(y|p)$. Note that the likelihood is proportional to probability. It is not the same as probability however, as probability should sum to one while the likelihood does not in general sum to one.

The objective is to determine the likelihood (or probability) of the model parameters given some observations, written as $L(p|y)$. This describes the distribution of the parameters. To obtain this likelihood, one uses Bayes' rule (hence Bayesian statistics):

$$L(p|y) = L(y|p) \cdot L(p)/L(y) \quad \text{(Bayes' Rule)}$$

with

$$L(y) = \int L(y|p) \cdot L(p) \quad \text{(Sum rule: sum over all plausible values for } p\text{)}$$

In this formula, $L(p)$ represents the so-called prior likelihood for the parameters, in short *prior*.

By means of this prior, one includes prior information, knowledge or beliefs about the parameter into the calculations. For example, if a certain parameter cannot be negative then one constructs a prior which is zero for negative values of the parameter: $L(p)_{p<0} = 0$. $L(y)$ represents the total likelihood of the data. This is the overall likelihood for the data to have been observed for all considered values for the parameters p. In Bayes' rule, $L(y)$ is a scaling factor which makes sure that $L(p|y)$ integrates to one and thus represents a probability. If one does not scale with $L(y)$ then one can still find the parameters which maximise $L(p|y)$ since $L(y)$ is a constant. Such parameters are called the maximum likelihood parameter estimates. However, to obtain confidence limits for the parameters, one relies on the probability and should scale properly. Hence, $L(y)$ is needed for the quantification of uncertainty. In general, the calculus of $L(y)$ is difficult because there is no closed form or analytic solution for this sum/integral equation. As a result, several methods have been developed to approximate this integral.

B.2 MONTE CARLO SIMULATION

In Monte Carlo methods, the uncertainties in the model inputs and parameters are expressed as probability distributions. Multivariate samples are then obtained using a statistical sampling method, propagated through the model using simulations, and the results are analysed to develop probability distributions for the model output variables.

With sufficient sampling from an unknown distribution, the true distribution can eventually be approximated numerically. This paradigm can be put to use in a classic frequency-based statistical framework, where the true parameters are considered fixed and distributions of parameter estimates are characterised, or in a Bayesian context, where the model parameters, are considered to be uncertain and where the distributions of the parameters, not their estimates, are characterised. In the latter, a prior distribution is set up for the parameters, which reflects the expected distribution for the parameters in the absence of experimental data and/or before experimentation (hence prior).

One of the earliest documented applications of the basic Monte Carlo method was in the determination of the value of π (Hall, 1873). The term 'Monte Carlo' was coined in the 1940s by researchers working on nuclear fusion at the Los Alamos National Laboratory (Metropolis & Ulam, 1949).

Monte Carlo methods generate the solution of the integral of the product of two variables. Many problems can be formulated in this context such as finding the mean of a stochastic variable which is defined as the integral of the variable multiplied by its probability density function. Monte Carlo methods are often used to evaluate difficult multidimensional integrals with complicated boundary conditions. The problem of estimating uncertainties in simulation results can be formulated as an integration problem. For example, the mean of the model outputs is the integral of the product of the model outputs and the joint probability density function.

The basic Monte Carlo method can require a large number of samples in order to converge. The uncertainty in Monte Carlo simulations is proportional to $1/\sqrt{n}$ (Eckhardt, 1987), where n is the number of samples. This means that every decimal point of extra accuracy requires 100 times the number of samples. As a result, Monte Carlo simulations could require hundreds or thousands of simulations to converge depending on the required accuracy.

In order to reduce the number of simulations that must be run, methods have been developed to generate more efficiently the sets of random numbers required as model inputs. These include Markov chain Monte Carlo (Metropolis *et al.*, 1953), stratified sampling methods such as Latin hypercube sampling (LHS) and quasi-Monte Carlo (see Torvi & Hertzberg, 1998). Quasi-Monte Carlo methods construct deterministic sequences such as the Halton, Sobol or Hammersley sequences that share properties of random or pseudo-random sequences. These methods are found to have less error than random Monte Carlo methods and require fewer samples to converge but the advantage may be slight in large problems (Morokoff & Caflisch, 1995).

The probability density functions used for the model input variables and parameters depend on the available data. In cases where data are available, the distribution of the data can be determined using statistical techniques. For variables for which little information is known except for expected minimum and maximum values, a uniform distribution is often used. A triangular distribution is used if a most likely value and minimum and maximum values are known.

B.2.1 Random sampling and LHS

In the random sampling (RS) procedure, at each Monte Carlo run, a vector of model parameters is randomly sampled from the joint distribution of parameters. The sampling of parameters at each Monte Carlo run is independent from the previous ones. Therefore, in this sampling approach, the coverage of the entire support of distributions (used for the characterisation of model parameter uncertainty) might not be guaranteed, unless a large-enough number of Monte Carlo simulations is performed.

An alternative sampling to the RS method for exploring the support of different parameter distributions is the LHS method. In the LHS method, the range of the input variable distribution is divided into N sub-intervals (e.g., $N = 4$ in Figure B.1) with equal probabilities. One value is selected from each sub-interval and this process is repeated for all the input variables. The generated input variable values are then paired randomly to generate a sequence of input samples for use in the Monte Carlo simulations. Compared to the RS method in which different samples are generated by RS directly from the entire range of distributions, in LHS, RS is performed in each sub-interval and all sub-intervals are sampled.

Figure B.1 illustrates the result of generating four vectors of parameters in a two-dimensional parameter space generated using RS and LHS methods. As indicated in (a), in this particular realisation of four samples, generated according to the RS method, no value is sampled from sub-interval (1) of parameter θ_1 and sub-interval (2) of parameter θ_2. However, the sampling result based on the LHS method indicates that the generated values include representatives from all sub-intervals.

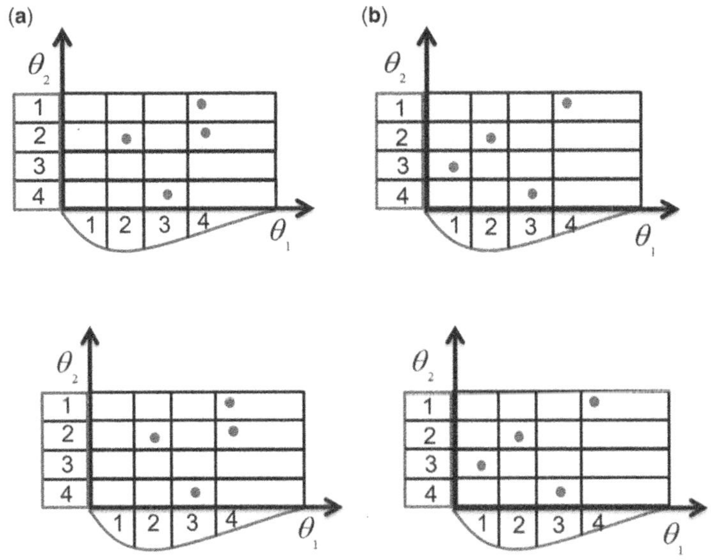

Figure B.1 (a) Schematic of a RS and (b) LHS procedures.

In general, the application of LHS could reduce the number of sampled values required to reach convergence of the output distributions (Tung & Yen, 2005). However, there could be some cases where LHS sampling would not lead to a more rapid convergence of output distribution compared to the RS sampling as the convergence also depends on the complexity of the model and its parameters.

In addition, in the RS method, the sampling of parameters at each Monte Carlo run is independent from the next one and in each run the convergence of the output distributions can be checked to determine whether more simulations are required or not. In contrast, in the LHS method, the number of Monte Carlo runs should be determined and samples generated before running any simulations. Therefore, if the selected number of Monte Carlo simulations turns out to be insufficient, the users cannot simply add more samples (like in the RS method) unless the consistency of the LHS procedure is insured.

A possible solution to increase the size of samples in the LHS method is the replicated LHS method (McKay *et al.*, 1979) in which instead of generating N number of samples using the LHS, k number of LHS designs with n number of samples each, is generated ($N = k \times n$). After the termination of each Monte Carlo simulation with n samples, the convergence of the output distributions is checked, and if more simulations are required, other n samples are generated using the LHS and Monte Carlo simulation continues using the newly generated samples. The efficiency of the repeated LHS depends on the appropriate choice of n as selecting it too large may not result in significant reductions of model runs and a value that is too small could result in inadequate coverage of the entire parameter space (Benedetti *et al.*, 2011).

B.2.2 Introducing correlations between parameters

One of the important factors in Monte Carlo simulations that could affect some of the statistical properties of the simulated output distributions is proper incorporation of possible correlation structures in the sampling of uncertain parameters. Different methods presented in the literature can be used to introduce a desired correlation structure among the sampled values (Iman & Conover, 1982; Tung & Yen, 2005). However, some of the methods suffer from certain shortcomings and their application depends on the validity of a set of assumptions regarding the marginal distribution of the parameters (Tung & Yen, 2005).

One of the commonly used methods for introducing correlation among uncertain parameters is the method of Iman–Conover (Iman & Conover, 1982). Being independent from the type of marginal distributions, applicability to any sampling scheme (e.g., RS or LHS), and relatively simple implementation have been mentioned as the main advantages of this method (Iman & Davenport, 1982).

REFERENCES

Benedetti L., Claeys F., Nopens I. and Vanrolleghem P. A. (2011). Assessing the convergence of LHS Monte Carlo simulations of wastewater treatment models. *Water Science Technology*, **63**(10), 2219–2224. http://hdl.handle.net/1854/LU-1922233 (accessed 2021).

Eckhardt R. (1987). Stan Ulam, John Von Neumann, and the Monte Carlo Method. *Los Alamos Science*, Special Issue.

Hall A. (1873). On an experimental determination of PI. *Messenger of Mathematics*, **2**, 113–114.

Iman R. and Conover W. J. (1982). A distribution-free approach to inducing rank correlation among input variables. *Communication in Statistics – Simulation and Computation*, **11**(3), 311–334.

Iman R. L. and Davenport J. M. (1982). An Iterative Algorithm to Produce a Positive Definite Correlation Matrix From an 'Approximate Correlation Matrix'. Sandia National Laboratories, Albuquerque, NM (USA), SAND81-1376. https://doi.org/10.2172/5152227 (accessed 2021).

Johnson R. A. and Wichern D. W. (2007). Applied Multivariate Statistical Analysis, 6th edn, Pearson Prentice Hall, Upper Saddle River, NJ.

McKay M. D., Beckman R. J. and Conover W. J. (1979). Comparison of three methods for selecting values of input variables in the analysis of output from a computer code. *Technometrics*, **21**(2), 239–245, doi: 10.1080/00401706.1979.10489755

Metropolis N. and Ulam S. (1949). The Monte Carlo method. *Journal of the American Statistical Association*, **44**, 335–341.

Metropolis N., Rosenbluth A. W., Rosenbluth M. N. and Teller A. H. (1953). Equation of state calculations by fast computing machines. *Journal of Chemical Physics*, **21**, 1087. https://doi.org/10.1063/1.1699114 (accessed 2021).

Morokoff W. J. and Caflisch R. E. (1995). Quasi-Monte Carlo integration. *Journal of Computational Physics*, **122**(2), 218–230.

Torvi H. and Hertzberg T. (1998). Methods of evaluating uncertainties in dynamic simulation – a comparison of performance. *Computers & Chemical Engineering*, **22**(Suppl), S985–S988.

Tung Y. K. and Yen B. C. (2005). Hydrosystem Engineering Uncertainty Analysis. McGraw-Hill Book Company, New York.

Appendix C

Existing methods for uncertainty analysis in WWT model-based projects – Complete literature search results

C.1 INTRODUCTION

Several literature searches were conducted to identify potential references for review. One of the authors conducted a search using ISI/Web of Knowledge (Science) in 2008 using various keywords related to wastewater treatment and uncertainty. In 2009 and 2010 the services of a research librarian were used to conduct additional literature searches. The searches of the research librarian included the data bases Compendex, Scopus, and Pollution Abstracts and Toxicology Abstracts 1998 to the present. These searches used key words and topics related to uncertainty in wastewater treatment (wastewater, treatment, uncertainty, evaluation, assessment, modelling, probabilistic, stochastic, sensitivity analysis) and also included searches using author and researcher names known to be working in the area.

The literature searches resulted in an initial list of over 500 references. Abstracts of the references in the initial list were reviewed and 386 references were determined to be directly relevant to the purpose of this chapter. The number of references in the short list fluctuated, with additions made during the detailed review process. An updated literature review for 2011–2019 has been added at the end of the chapter.

While the list cannot be considered comprehensive or complete, it does represent a credible search of the literature and provides the current status of literature activity regarding the subject. The categorization that follows is based on the judgement of the authors and hence is not unique. For example, there are some references assigned to one category that address several other categories. Similarly, some reference studies maybe listed in a category which are not perfectly explained by the concise category titles itself. However, the present choice of categories serves as a framework to synthesize the considerable breadth of the topic and size of the literature search results into discrete and discussable topics.

Editor(s): Evangelia Belia, Lorenzo Benedetti, Bruce Johnson, Sudhir Murthy, Marc Neumann, Peter Vanrolleghem and Stefan Weijers
doi: 10.2166/9781780401034_0151

C.2 ASSESSMENT OF INPUT AND PARAMETER UNCERTAINTY

Angelbeck D. I. and Minkara R. Y. (1994). Strange attractors and chaos in waste-water flow. *Journal of Environmental Engineering-ASCE*, **120**(1), 122–137.

Angelbeck D. I. and Minkara R. Y. (1996). Strange attractors and chaos in wastewater flow – closure. *Journal of Environmental Engineering-ASCE*, **122**(3), 242–243.

Bertrand-Krajewski J.-L., Winkler S., Saracevic E., Torres A. and Schaar H. (2007). Comparison of and uncertainties in raw sewage COD measurements by laboratory techniques and field UV-visible spectrometry. *Water Science and Technology*, **56**(11), 17–25.

Bormann N. E. and Kincanon E. (1996). Strange attractors and chaos in wastewater flow – discussion. *Journal of Environmental Engineering-ASCE*, **122**(3), 240–240.

Checchi N. and Marsili-Libelli S. (2005). Reliability of parameter estimation in respirometric models. *Water Research*, **39**(15), 3686–3696.

Cierkens K., Plano S., Benedetti L., Weijers S., de Jonge J. and Nopens I. (2012). Impact of influent data frequency and model structure on the quality of WWTP model calibration and uncertainty. *Water Science and Technology*, **65**(2), 233–242.

Cox C. D. (2004). Statistical distributions of uncertainty and variability in activated sludge model parameters. *Water Environment Research*, **76**(7), 2672–2685.

Daebel H., Manser R. and Gujer W. (2007). Exploring temporal variations of oxygen saturation constants of nitrifying bacteria. *Water Research*, **41**(5), 1094–1102.

Drolc A. and Ro M. (2002). Uncertainty of ammonium nitrogen determination in wastewater. First International Conference on Waste Management and the Environment, Waste Management 2002, September 4, 2002–September 6, 2002, Cadiz, Spain, Wit Press.

Drolc A. and Ro M. (2004). Assuring comparability of results of nitrate determinations in wastewater: Application of metrological principles. Second International Conference on Waste Management and the Environment, September 29, 2004–October 1, 2004, Rhodes, Greece, WITPress.

Fang F., Ni B. and Yu H. G. (2009). Estimating the kinetic parameters of activated sludge storage using weighted non-linear least-squares and accelerating genetic algorithm. *Water Research*, **43**(10), 2595–2604.

Friedler E. and Butler D. (1996). Quantifying the inherent uncertainty in the quantity and quality of domestic wastewater. *Water Science and Technology*, **33**(2), 65–78.

Guimaraes A. D., Carvalho J. J., Gonçalves C. and Alpendurada M. F. (2008). Simultaneous analysis of 23 priority volatile compounds in water by solid-phase microextraction-gas chromatography-mass spectrometry and estimation of the method's uncertainty. *International Journal of Environmental Analytical Chemistry*, **88**(3), 151–164.

Huo J. (2005). Application of statistical methods and process models for the design and analysis of activated sludge wastewater treatment plants (WWTPs), The University of Tennessee.

Huo J., Cox C., Seaver W., Robinson B. and Jiang Y. (2008). Innovative Missing Data Replacement Methods Using Time Series Models, ASCE, Honolulu, Hawaii.

Hvala N., Zec M. and Strmcnik S. (2008). Non-linear model parameter estimation – estimating a feasible parameter set with respect to model use. *Mathematical and Computer Modelling of Dynamical Systems*, **14**(6), 587–605.

Joannis C., Ruban G., Gromaire M.-C., Cheboo G. and Bertrand-Krajewski J.-L. (2008). Reproducibility and uncertainty of wastewater turbidity measurements. *Water Science and Technology*, **57**(10), 1667–1673.

Kim H. S., Yoon Y. N. and Kim J. H. (2001). Searching for strange attractor in wastewater flow. *Stochastic Environmental Research and Risk Assessment*, **15**(5), 399–413.

Leander R., Buishand A., Aladers P. and De Wit M. (2005). Estimation of extreme floods of the River Meuse using a stochastic weather generator and a rainfall-runoff model. *Hydrological Sciences Journal*, **50**(6), 1089–1104.

Liu X.-P. and Zhou Y.-W. (2005). Stochastic simulation of sanitary flow series using Markov chain. *Harbin Gongye Daxue Xuebao/Journal of Harbin Institute of Technology*, **37**(5), 642–645.

Lockie T., Cowpertwait P., Davis M., Kinley P. and Paterson G. (2003). Evaluation of Historic and Stochastic Rainfall in Network Modelling in the Integrated Catchment Study of Auckland, New Zealand. World Water and

Environmental Resources Congress 2003, June 23, 2003–June 26, 2003, Philadelphia, PA, United States, American Society of Civil Engineers.

Marsili-Libelli S., D'Ardes V. and Bondi C. (2008). Development of a open-vessel single-stage respirometer. *Water Science and Technology*, **57**(10), 1675–1682.

Marsili-Libelli S. and Giunti L. (2002). Fuzzy predictive control for nitrogen removal in biological wastewater treatment. *Water Science and Technology*, **45**(4–5), 37–44.

Ort C., Schaffner C., Giger W. and Gujer W. (2005). Modelling stochastic load variations in sewer systems. *Water Science and Technology*, **52**(5), 113–122.

Panjan J., Bogataj M. and Kompare B. (2005). Statistical analysis of the equivalent design rainfall. *StatistiÄna Analiza Gospodarsko Enakovrednih Nalivov Strojniski Vestnik/Journal of Mechanical Engineering*, **51**(9), 600–611.

Puig S., van Loosdrecht M., Colprim J. and Meijer S. C. F. (2008). Data evaluation of full-scale wastewater treatment plants by mass balance. *Water Research*, **42**(18), 4645–4655.

Rauch W., Brockmann D., Peters I., Larsen T. A. and Gujer W. (2003). Combining urine separation with waste design: An analysis using a stochastic model for urine production. *Water Research*, **37**(3), 681–689.

Rieger L., Thomann M., Gujer W. and Siegrist H. (2005). Quantifying the uncertainty of on-line sensors at WWTPs during field operation. *Water Research*, **39**(20), 5162–5174.

Rieger L., Langergraber G. and Siegrist H. (2006). Uncertainties of spectral in situ measurements in wastewater using different calibration approaches. *Water Science and Technology*, **53**(12), 187–197.

Robinson R. B., Cox C. D. and Odom K. (2005). Identifying outliers in correlated water quality data. *Journal of Environmental Engineering*, **131**(4), 651–657.

Ruban G. and Joannis C. (2008). Evaluation of calibration curve uncertainty using the Monte Carlo method Application to turbidity measurement. Évaluation de l'incertitude des courbes d'étalonnage par la méthode de Monte Carlo. Application a la turbidimétrie. *Bulletin des Laboratoires des Ponts et Chaussees*, (272), 33–43.

Thomann M. (2008). Quality evaluation methods for wastewater treatment plant data. *Water Science and Technology*, **57**(10), 1601–1609.

Tsai C. and Franceschini S. (2005). Evaluation of probabilistic point estimate methods in uncertainty analysis for environmental engineering applications. *Journal of Environmental Engineering*, **131**, 387.

Winkler S., Bertrand-Krajewski J.-L., Saracevicc E. and Torres A. (2008). Benefits, limitations and uncertainty of in situ spectrometry. *Water Science and Technology*, **57**(10), 1651–1658.

Young P. L. and Komisar S. J. (1999). The variability introduced by partial sample analysis to numbers of Cryptosporidium oocysts and Giardia cysts reported under the information collection rule. *Water Research*, **33** (11), 2660–2668.

C.3 ASSESSMENT OF MODEL STRUCTURE UNCERTAINTY

Abusam A. and Keesman K. (2002). A sensitivity analysis of the secondary settling tank double-exponential model. European Water Management [Online]. pp. 1–9. Available: http://www.ewaonline.de/journal/2002_07.pdf (accessed August 03 2021). European Water Management Online Official Publication of the European Water Association (EWA) © EWA 2002..

Beck M. B. (1999). Coping with ever larger problems, models, and data bases. *Water Science and Technology*, **39**(4), 1–11.

Blind M. W. and Refsgaard J. C. (2007). Operationalising uncertainty in data and models for integrated water resources management. *Water Science and Technology*, **56**(9), 1–12.

Brugnach M., Tagg A., Keil F. and de Lange W. J. (2007). Uncertainty matters: computer models at the science-policy interface. *Methods for Participatory Water Resources Management*, **21**(7), 1075–1090.

Cierkens K., Plano S., Benedetti L., Weijers S., de Jonge J. and Nopens I. (2012). Impact of influent data frequency and model structure on the quality of WWTP model calibration and uncertainty. *Water Science and Technology*, **65**(2), 233–242.

Claeys P., van Griensven A., Benedetti L., De Baets B. and Vanrolleghem P. A. (2010). Conceptual thinking on numerical solver selection and uncertainty terminology. *Journal of Hydroinformatics*, **12**(3), 241–250.

Curlin M., Matic M., Matosic M., Mijatovic I. and Kurtanjek Z. (2004). Effects of hydraulic residence time and mixing on wastewater treatment in a membrane bioreactor. *Chemical and Biochemical Engineering Quarterly*, **18**(2), 97–104.

De Clercq B., Coen F., Vanderhaegen B. and Vanrolleghem P. A. (1999). Calibrating simple models for mixing and flow propagation in waste water treatment plants. Proceedings of the 1998 Aquatech 1998 Conference on Application of Models in Water Management, September 24, 1998–September 25, 1998, Amsterdam, NLD, Elsevier Ltd.

Dette H., Melas V., Pepelyshev A. and Strigul N. (2003). Efficient design of experiments in the Monod model. *Journal of the Royal Statistical Society, Series B, Statistical Methodology*, **65**(3), 725–742.

Dette H., Melas V., Pepelyshev A. and Strigul N. (2005). Robust and efficient design of experiments for the Monod model. *Journal of Theoretical Biology*, **234**(4), 537–550.

Dochain D., Vanrolleghem P. A. and Van Daele M. (1995). Structural identifiability of biokinetic models of activated sludge respiration. *Water Research*, **29**(11), 2571–2578.

Gernaey K. V., Van Loosdrecht M. C. M., Henze M., Lind M. and Jorgensen S. (2004). Activated sludge wastewater treatment plant modelling and simulation: state of the art. *Environmental Sciences and Artificial Intelligence Environmental Modelling and Software*, **19**(9), 763–783.

Graham D., Knapp C., Van Vleck E. S., Bloor K., Lane T. B. and Graham C. E. (2007). Experimental demonstration of chaotic instability in biological nitrification. *The ISME Journal*, **1**(5), 385–393.

Grau P., de Gracia M., Vanrolleghem P. A. and Ayesa E. (2007). A new plant-wide modelling methodology for WWTPs. *Water Research*, **41**(19), 4357–4372.

Guisasola A., Sin G., Baeza J. A., Carrera J. and Vanrolleghem P. A. (2005). Limitations of ASM1 and ASMS3: A comparison based on batch oxygen uptake rate profiles from different full-scale wastewater treatment plants. *Water Science and Technology*, **52**(10–11), 69–77.

Gujer W. (2002). Microscopic versus macroscopic biomass models in activated sludge systems. *Water Science and Technology*, **45**(6), 1–11.

Haider S., Svardal K., Vanrolleghem P. A. and Kroiss H. (2003). The effect of low sludge age on wastewater fractionation (SS, SI). *Water Science and Technology*, **47**(11), 203–209.

Ibrahim G., Ramadan M., El-Marouf S. A. A. and Al-Hahdi A. M. (2007). Periodic and non periodic (complex) behavior of a model of bioreactor with cell recycling. *Journal of Mathematics and Statistics*, **3**(1), 1–11.

Ibrahim G., Habib H. and Saleh O. (2008). Periodic and chaotic solutions for a model of a bioreactor with cell recycle. *Biochemical Engineering Journal*, **38**(2), 124–137.

Insel G., Karahan Ozgun O., Orhen D. and Vanrolleghem P. A. (2002). Important limitations in the modeling of activated sludge: biased calibration of the hydrolysis process. *Water Science and Technology*, **45**(12), 23–36.

Kops S. and Vanrolleghem P. (1996). An Evaluation of Methodologies for Uncertainty Analysis in Biological Waste Water Treatment. BIOMATH, Ghent University, Belgium.

Kops S., Vangheluwe H., Claeys F., Vanrollghem P., Yuan Z. and Vansteenkiste G. (1999). The process of model building and simulation of ill-defined systems: application to wastewater treatment. *Mathematical and Computer Modelling of Dynamical Systems*, **5**(4), 298–312.

Krasnochtanova I. A. R., Kletting M., Aschemann H., Hofer E. P. and Schoop K.-M. (2009). Interval methods as a simulation tool for the dynamics of biological wastewater treatment processes with parameter uncertainties. *Applied Mathematical Modelling*, **34**, 744–762.

Kumar S., Kumar M., Thurow K., Stoll R. and Kragl U. (2008). Fuzzy filtering for robust bioconcentration factor modelling. *Environmental Modelling and Software*, **24**(1), 44–53.

Lavallee B., Lessard P. and Vanrolleghem P. A. (2005). Modeling the metabolic adaptations of the biomass under rapid growth and starvation conditions in the activated sludge process. *Journal of Environmental Engineering and Science*, **4**(6), 533–548.

Lee T. T., Wang F. Y. and Newell R. B. (1999a). Distributed parameter approach to the dynamics of complex biological processes. *AIChE Journal*, **45**(10), 2245–2261.

Lee T. T., Wang F. Y. and Newell R. B. (1999b). On the modelling and simulation of a BNR activated sludge process based on distributed parameter approach. *Water Science and Technology*, **39**(6), 79–88.

Linke P. and Kokossis A. (2004). Advanced process systems design technology for pollution prevention and waste treatment. *Advances in Environmental Research*, **8**(2), 229–245.

Meng D.-T. and Qing X.-X. (2008). Applications and development of expert system in wastewater treatment systems. *Chongqing Jianzhu Daxue Xuebao/Journal of Chongqing Jianzhu University*, **30**(1), 146–149.

Moisan M., Bernard O. and Gouze J.-L. (2009). Near optimal interval observers bundle for uncertain bioreactors. *Automatica*, **45**(1), 291–295.

Neumann M. B. and Gujer W. (2008). Underestimation of uncertainty in statistical regression of environmental models: influence of model structure uncertainty. *Environmental Science and Technology*, **42**(11), 4037–4043.

Omlin M. and Reichert P. (1999). A comparison of techniques for the estimation of model prediction uncertainty. *Ecological Modelling*, **115**(1), 45–59.

Paladino O. (1994). Transition to chaos in continuous-processes – applications to waste-water treatment reactors. *Environmetrics*, **5**(1), 57–70.

Paladino O., Ratto M. and Costa P. (1995). Chaos and chemical reactor models: sensitivity of dynamics on parameters uncertainty. *Chemical Engineering Science*, **50**(23), 3829–3833.

Plosz B., De Clercq J., Nopens I., Benedetti L. and Vanrolleghem P. A. (2011). Shall we upgrade one-dimensional secondary settler models used in WWTP simulators? – An assessment of model structure uncertainty and its propagation. *Water Science and Technology*, **63**(8), 1726–1738.

Raduly B., Gernaey K. V., Capodaglio A. G., Mikkelsen P. S. and Henze M. (2007). Artificial neural networks for rapid WWTP performance evaluation: methodology and case study. *Environmental Modelling and Software*, **22**(8), 1208–1216.

Rauh A., Kletting M., Aschemann H. and Hofer E. P. (2004). Application of interval arithmetic simulation techniques to wastewater treatment processes. Proceedings of the IASTED International Conference on Modelling. Identification and Control, 23–25 February 2004, Grindelwald, Switzerland.

Rauh A., Kletting M., Aschemann H. and Hofer E. P. (2005). Robust controller design for bounded state and control variables and uncertain parameters using interval methods. Proceedings of the 5th International Conference on Control and Automation (CCA), 26–29 June 2005, Budapest, Hungary.

Rauh A., Kletting M., Aschemann H. and Hofer E. P. (2007). Reduction of overestimation in interval arithmetic simulation of biological wastewater treatment processes. *Journal of Computational and Applied Mathematics*, **199**(2), 207–212.

Rustum R., Adeloye A. and Simala A. (2007). Kohonen self-organising map (KSOM) extracted features for enhancing MLP-ANN prediction models of BOD5. International Symposium: Water Quality and Sediment Behaviour of the Future: Predictions for the 21st Century – 24th General Assembly of the International Union of Geodesy and Geophysics (IUGG), July 11, 2007–July 13, 2007, Perugia, Italy, IAHS Press.

Sorensen P. B., Lerche D. B. and Bruggemann R. (2005). Resources and uncertainties in evaluation of chemicals: basic methodological developments. *Water Science Technology*, **52**(6), 235–242.

Schuler A. (2005). Diversity matters: dynamic simulation of distributed bacterial states in suspended growth biological wastewater treatment systems. *Biotechnology and Bioengineering*, **91**, 1.

Schuler A. (2006). Process hydraulics, distributed bacterial states, and biological phosphorus removal from wastewater. *Biotechnology and Bioengineering*, **94**, 5.

Schuler A. J. and Jenkins D. (2003). Enhanced biological phosphorus removal from wastewater by biomass with different phosphorus contents, part I: experimental results and comparison with metabolic models. *Water Environment Research*, **75**(6), 485–498.

Schuler A. J., Jenkins D. and Ronen P. (2001). Microbial storage products, biomass density, and settling properties of enhanced biological phosphorus removal activated sludge. *Water Science and Technology*, **43**(1), 173–180.

Sin G., Gernaey K. V., Neumann M. B., van Loosdrecht M. C. M. and Gujer W. (2009). Uncertainty analysis in WWTP model applications: a critical discussion using an example from design. *Water Research*, **43**(11), 2894–2906.

Sin G. and Vanrolleghem P. A. (2006). Evolution of an ASM2s-like model structure due to operational changes of an SBR process. *Water Science and Technology*, **53**(12), 237–245.

Sin G., Van Hulle S. W. H., DePauw D. J. W., van Griensven A. and Vanrolleghem P. A. (2005). A critical comparison of systematic calibration protocols for activated sludge models: A SWOT analysis. *Water Research*, **39**(12), 2459–2474.

Sorensen P. B., Lerche D. B. and Bruggemann R. (2005). Resources and uncertainties in evaluation of chemicals: basic methodological developments. *Water Science and Technology*, **52**(6), 235–242.

Strigul N., Dette H. and Melas V. B. (2009). A practical guide for optimal designs of experiments in the Monod model. *Environmental Modelling and Software*, **24**(9), 1019–1026.

Stroot P., Saikaly P. and Oerther D. B. (2005). Dynamic growth rates of microbial populations in activated sludge systems. *Journal of Environmental Engineering*, **131**, 1698.

Su Y.-M., Makinia J. and Pagilla K. R. (2008). Estimation of autotrophic maximum specific growth rate constant – experience from the long-term operation of a laboratory-scale sequencing batch reactor system. *Water Environment Research*, **80**(4), 355–366.

Tabatabaei T. S., Farooq T., Guergachi A. and Krishnan S. (2007). Support vector machines-based approach for chemical phosphorus removal process in wastewater treatment plant. 2006 Canadian Conference on Electrical and Computer Engineering, CCECE'06, May 7, 2006–May 10, 2006, Ottawa, ON, Canada, Institute of Electrical and Electronics Engineers Inc.

Tenno R. and Uronen P. (1995). State and parameter estimation for wastewater treatment processes using a stochastic model. *Control Engineering Practice*, **3**(6), 793–804.

Tsai J. C. C. and Chen V. C. P. (2005). Flexible and robust implementations of multivariate adaptive regression splines within a wastewater treatment stochastic dynamic program. *Quality and Reliability Engineering International*, **21**(7), 689–699.

Vanrolleghem P. A. and Keesman K. J. (1996). Identification of biodegradation models under model and data uncertainty. Proceedings of the 1995 IAWQ Interdisciplinary Symposium on Uncertainty, Risk and Transient Pollution Events, July 26, 1995–July 28, 1995, Exeter, UK, Pergamon Press Inc.

Von Sperling M. (2002). Relationship between first-order decay coefficients in ponds, for plug flow, CSTR and dispersed flow regimes. *Water Science and Technology*, **45**(1), 17–24.

Wan J.-Q., Huang M.-Z. and Ma Y.-W. (2008). Studies on predicting the effluent treatment process with fuzzy neural network. *Zhongguo Zaozhi Xuebao/Transactions of China Pulp and Paper*, **23**(2), 96–99.

Whiteley C. G. and Lee D.-J. (2006). Enzyme technology and biological remediation. *Enzyme and Microbial Technology*, **38**(3–4), 291–316.

Zhang Y. and Henson M. (2001). Bifurcation analysis of continuous biochemical reactor models. *Biotechnology Progress*, **17**(4), 647–660.

Zhang F.-F., Huang A.-Q., Xie S.-q. and He J. (2008). Modeling and simulation of adsorption and biodegradation in activated sludge process based on cellular automata. IEEE International Conference on Automation and Logistics, ICAL 2008, September 1, 2008–September 3, 2008, Qingdao, China, Inst. of Elec. and Elec. Eng. Computer Society.

Zlateva P. and Dimitrova N. (2005). Stability analysis of a nonlinear model of wastewater treatment processes. Third International Conference on Numerical Analysis and Its Applications, NAA 2004, June 29, 2004–July 3, 2004, Rousse, Bulgaria, Springer Verlag.

C.4 PROPAGATION OF UNCERTAINTY FOR MODEL-BASED DECISIONS

Abraham D. M. (2003). Life Cycle Cost Integration for the Rehabilitation of Wastewater Infrastructure. Construction Research Congress, Winds of Change: Integration and Innovation in Construction, Proceedings of the Congress, March 19, 2003–March 21, 2003, Honolulu, HI, United states, American Society of Civil Engineers.

Abusam A., Keesman KJ. ., van Straten G., Spanjers H. and Meinema K. (2000). Parameter estimation procedure for complex non-linear systems: calibration of ASM No. 1 for N-removal in a full-scale oxidation ditch. *WATERMAX* , **43**(7), 357–365.

Abusam A., Keesman KJ. ., van Straten G., Spanjers H. and Meinema K. (2002). Estimation of uncertainties in the performance indices of an oxidation ditch benchmark. *Journal of Chemical Technology and Biotechnology*, **77** (9), 1058–1067.

Abusam A., Keesman K. J. and van Straten G. (2003). Forward and backward uncertainty propagation: an oxidation ditch modelling example. *Water Research*, **37**(2), 429–435.

Afonso P. M. and da Conceicao Cunha M. (2007). Robust optimal design of activated sludge bioreactors. *Journal of Environmental Engineering*, **133**(1), 44–52.

Al-Redhwan S. A. and Crittenden B. D. and Lababidi, H.M.S. (2005). Wastewater minimization under uncertain operational conditions. *Computers and Chemical Engineering*, **29**(5), 1009–1021.

Arboleda C. A. and Abraham D. M. (2006). Evaluation of flexibility in capital investments of infrastructure systems. *Engineering, Construction and Architectural Management*, **13**(3), 254–274.

Banda C. G., Sleigh P. A. and Mara D. D. (2005). *Escherichia coli* removal in waste stabilization ponds: a comparison of modern and classical designs. *Water Science and Technology*, **51**(12), 75–81.

Batstone D. J., Torrijos M., Ruiz C. and Schmidt J. E. (2004). Use of an anaerobic sequencing batch reactor for parameter estimation in modelling of anaerobic digestion. *Water Science and Technology*, **50**(10), 295–303.

Bechmann H., Nielsen M. K., Madsen H. and Poulsen N. K. (1999). Grey-box modelling of pollutant loads from a sewer system. *Urban Water*, **1**(1), 71–78.

Bechmann H., Madsen H., Poulsen N. and Nielsen M. K. (2000). Grey box modeling of first flush and incoming wastewater at a wastewater treatment plant. *Environmetrics*, **11**(1), 1–12.

Benedetti L., Bixio D. and Vanrolleghem P. A. (2006a). Benchmarking of WWTP design by assessing costs, effluent quality and process variability. *Water Science and Technology*, **54**(10), 95–102.

Benedetti L., Bixio D. and Vanrolleghem P. A. (2006b). Assessment of WWTP design and upgrade options: balancing costs and risks of standards' exceedance. *Water Science and Technology*, **54**(6–7), 371–378.

Benedetti L., Bixio D., Claeys F. and Vanrolleghem P. A. (2008). Tools to support a model-based methodology for emission/immission and benefit/cost/risk analysis of wastewater treatment systems which considers uncertainties. *Environmental Modelling & Software*, **23**(8), 1082–1091.

Benedetti L., De Keyser W., Nopens I. and Vanrolleghem P. A. (2010). Probabilistic modelling and evaluation of waste water treatment plant upgrades in the EU Water Framework Directive context. *Journal of Hydroinformatics*, **12**(4), 380–395.

Benedetti L., Claeys F., Nopens I. and Vanrolleghem P. A. (2011). Assessing the convergence of LHS Monte Carlo simulations of wastewater treatment models. *Water Science and Technology*, **63**(10), 2219–2224.

Berthouex P. and Polkowski L. (1970). Optimum waste treatment plant design under uncertainty. *Journal (Water Pollution Control Federation)*, **42**(9), 1589–1613.

Bixio D., Carrette R. and Thoeye C. (2002). Two years of full-scale storm operation experience. *Water Science and Technology*, **46**(6–7), 167–172.

Boccelli D., Small M. and Diwekar U. (2007). Drinking water treatment plant design incorporating variability and uncertainty. *Journal of Environmental Engineering*, **133**(3), 303–312.

Chang N. B. and Hernandez E. A. (2008). Optimal expansion strategy for a sewer system under uncertainty. *Environmental Modeling & Assessment*, **13**(1), 93–113.

Chang N.-B., Chen W. C. and Shieh W. K. (2001). Optimal control of wastewater treatment plants via integrated neural network and genetic algorithms. *Civil Engineering and Environmental Systems*, **18**(1), 1–17.

Chang N.-B., Hernandez E. A and Davila E. (2005). Optimal risk-based expansion strategy of sewer infrastructure system in a fast growing Urban region under uncertainty. 2005 World Water and Environmental Resources Congress, May 15, 2005–May 19, 2005, Anchorage, AK, United States, American Society of Civil Engineers.

Chen H. W. and Chang N. B. (2002). A comparative analysis of methods to represent uncertainty in estimating the cost of constructing wastewater treatment plants. *Journal of Environmental Management*, **65**(4), 383–409.

Chen H.-W. and Chang N.-B. (2006). Decision support for allocation of watershed pollution load using grey fuzzy multiobjective programming. *Journal of the American Water Resources Association*, **42**(3), 725–745.

Chen C.-F., Chen Y.-C. and Lin J.-Y. (2008). Determination of optimal water resource management through a fuzzy multiobjective programming and genetic algorithm: case study in Kinman, Taiwan. *Practice Periodical of Hazardous,* Toxic, and Radioactive Waste Management, **12**(2), 86–95.

Chiu K., Lyn D. A., Savoye P. and Blatchley III, E. R. (1999). Integrated UV disinfection model based on particle tracking. *Journal of Environmental Engineering*, **125**(1), 7–16.

Chowdhury S., Champagne P. and Sarkar A. (2007). Use of treated wastewater: Evaluation of wastewater minimization strategies using fuzzy techniques. 2nd IASTED International Conference on Advanced Technology in the Environmental Field, ATEF 2006, Febrary 6, 2006 – Febrary 8, 2006, Lanzarote, Canary Islands, Spain, Acta Press.

Chung G. and Lansey K. (2009). Application of the shuffled frog leaping algorithm for the optimization of a general large-scale water supply system. *Water Resources Management*, **23**(4), 797–823.

Coveney M. F., Lowe E. F. and Battoe L. E. (2001). Performance of a recirculating wetland filter designed to remove particulate phosphorus for restoration of Lake Apopka (Florida, USA). *Water Science Technology*, **44**(11–12), 131–136.

Curtis T. P. and Sloan W. T. (2006). Towards the design of diversity: stochastic models for community assembly in wastewater treatment plants. *Water Science and Technology*, **54**(1), 227–236.

Del Grosso A. and Lanata F. (2006). The role of monitoring in the management of uncertainties and residual life of existing structures. 3rd International Conference on Bridge Maintenance, Safety and Management – Bridge Maintenance, Safety, Management, Life-Cycle Performance and Cost, July 16, 2006–July 19, 2006, Porto, Portugal, Taylor & Francis/Balkema.

Doby T. A., Loughlin D. H., de los Reyes III, F. L. and Ducoste J. J. (2002). Optimization of activated sludge designs using genetic algorithms. *Water Science and Technology*, **45**(6), 187–198.

Dochain D., Agathos S. N. and Vanrolleghem P. (1997). Asymptotic observers as a tool for modelling process dynamics. *Water Science and Technology*, **36**(5), 259–268.

Dominguez D. and Gujer W. (2006). Evolution of a wastewater treatment plant challenges traditional design concepts. *Water Research*, **40**(7), 1389–1396.

Dong H.-G., Lin C.-Y. and Chang C.-T. (2008). Simultaneous optimization approach for integrated water-allocation and heat-exchange networks. *Chemical Engineering Science*, **63**(14), 3664–3678.

Duchesne S. and Villeneuve J.-P. (2006). Estimation of drinking water production total costs: the Quebec City case study Estimation du cout total associe a la production d'eau potable: Cas d'application de la ville de Quebec. *Revue des Sciences de L'Eau*, **19**(2), 69–85.

Duchesne S., Beck M. B. and Reda A. (2001). Ranking stormwater control strategies under uncertainty: the river Cam case study. *Water Science and Technology*, **43**(7), 311–320.

Dyson B. E. (2006). Design framework for sizing free water surface wetlands subject to intermittent wastewater loading in aquaculture facilities. *Dissertation Abstracts International*, **68**(1, Suppl. B), 162.

El-Din A. G. and Smith D. W. (2002). A combined transfer-function noise model to predict the dynamic behavior of a full-scale primary sedimentation tank. *Water Research*, **36**(15), 3747–3764.

Elshorbagy W. E., Yakowitz D. S. and Lansey KE.. (1997). Design of engineering systems using stochastic decomposition: water supply planning application. *Engineering Optimization*, **27**(4), 279–302.

Flores-Alsina X. (2008). Conceptual design of wastewater treatment plants using multiple objectives, PhD thesis, Universitat de Girona, Girona, Spain, ISBN: 978-84-691-5689-6

Franks S. W. (1999). Disaggregation of environmental factors affecting sewer pipe failures. *Journal of Infrastructure Systems*, **5**(4), 150–158.

Freni G., Mannina G. and Vivani G. (2008). Uncertainty in urban stormwater quality modelling: the effect of acceptability threshold in the GLUE methodology. *Water Research*, **42**(8–9), 2061–2072.

Freni G., Mannina G. and Vivani G. (2009a). Uncertainty assessment of an integrated urban drainage model. *Journal of Hydrology*, **373**(3–4), 392–404.

Freni G., Mannina G. and Vivani G. (2009b). Urban runoff modelling uncertainty: comparison among Bayesian and pseudo-Bayesian methods. *Environmental Modelling and Software*, **24**(9), 1100–1111.

Friedler E., Shwartzman Z. and Ostfeld A. (2008). Assessment of the reliability of an on-site MBR system for greywater treatment and the associated aesthetic and health risks. *Water Science and Technology*, **57**(7), 1103–1110.

Fuentes M., Mussati M. C., Scenna N. J. and Aguirre P. A. (2009). Global modeling and simulation of a three-phase fluidized bed bioreactor. *Computers and Chemical Engineering*, **33**(1), 359–370.

Garcia-Salas J. C. and Chocat B. (2006). Assessment of uncertainties in the modelling of CSOs. *Water Science and Technology*, **54**(6–7), 247–254.

Griborio A., McCorquodale J., Pitt P., Vinci P. and Wang T.(2007). Uncertainty analysis in wastewater engineering modeling: understanding the limitations. *Proceedings of the Water Environment Federation* **2007**(17), 1584–1596.

Grismer M. E. (2005). Simulation evaluation of the effects of non-uniform flow and degradation parameter uncertainty on subsurface-flow constructed wetland performance. *Water Environment Research*, **77**(6), 3047–3053.

Guo P., Huang G. H. and He L. (2008). ISMISIP: an inexact stochastic mixed integer linear semi-infinite programming approach for solid waste management and planning under uncertainty. *Stochastic Environmental Research and Risk Assessment*, **22**(6), 759–775.

Gupta A. K. and Shrivastava R. K. (2006). Uncertainty analysis of conventional water treatment plant design for suspended solids removal. *Journal of Environmental Engineering*, **132**(11), 1413–1421.

Hahn M. A., Palmer R. N., Merrill MS. . and Lukas AB.. (2002). Expert system for prioritizing the inspection of sewers: knowledge base formulation and evaluation. *Journal of Water Resources Planning and Management*, **128**(2), 121–129.

Ham J. H., Yoon C. G., Jung KW. and Jang JH... (2007). Integrated modelling under uncertainty in watershed-level assessment and management. *Water Science and Technology*, **56**(8), 31–39.

Hauger M. B., Rauch W., Linde JJ. . and Mikkelsen PS.. (2002). Cost benefit risk – A concept for management of integrated urban wastewater systems? *Water Science and Technology*, **45**(3), 185–193.

He L., Huang G., Zeng G. and Lu H. (2008). Fuzzy inexact mixed-integer semiinfinite programming for municipal solid waste management planning. *Journal of Environmental Engineering*, **134**(7), 572–581.

Hilgert N., Harmand J., Steyer J.-P. and Vila J.-P. (2000). Nonparametric identification and adaptive control of an anaerobic fluidized bed digester. *Control Engineering Practice*, **8**(4), 367–376.

Huo J. (2005). Application of Statistical Methods and Process Models for the Design and Analysis of Activated Sludge Wastewater Treatment Plants (WWTPs). The University of Tennessee, Knoxvile, TN, USA.

Huo J., Jiang Y., Seaver W. L. and Robinson R. B. (2006). Statistically Based Design of Wastewater Treatment Plants (WWTPs) Using Monte Carlo Simulation of Activated Sludge Model No.1 (ASM1), Omaha, Nebraska, USA, ASCE.

Jiang T., Liu X., Kennedy M. D., Schippers J. C. and Vanrolleghem P. A. (2005). Calibrating a side-stream membrane bioreactor using activated sludge model no. 1. *Water Science and Technology*, **52**(10–11), 359–367.

Jiang R., Zeng G.-M., Li X, Du C., Li J. and Song L. (2007). Wastewater treatment process design optimization based on set pair analysis. *Hunan Daxue Xuebao/Journal of Hunan University Natural Sciences*, **34**(11), 70–75.

Kehl O., Wichern M., Lubken M. and Horn H. (2009). Analysis of design approaches for stabilization ponds under different boundary conditions – a comparison. *Ecological Engineering*, **35**(8), 1117–1128.

Kieffer M. and Walter E. (2004). Guaranteed nonlinear state estimator for cooperative systems. *Numerical Algorithms*, **37**(1–4 SPEC. ISS.), 187–198.

Kletting M., Rauh A., Aschemann H. and Hofer E. P. (2007). Consistency tests in guaranteed simulation of nonlinear uncertain systems with application to an activated sludge process. *Journal of Computational and Applied Mathematics*, **199**(2), 213–219.

Koch G., Kahni M. and Siergrist H. (2001). Calibration and validation of an ASM3-based steady-state model for activated sludge systems – part I: prediction of nitrogen removal and sludge production. *Water Research*, **35** (9), 2235–2245.

Kops S., Gernaey K., Thas O. and Vanrolleghem P. A. The Modelling of Noise Processes in Stochastic Differential Equations: Application To Biotechnological Processes. BIOMATH, University of Gent, Belgium.

Lauchlan C., Forty J. and May R. (2005). Flow resistance of wastewater pumping mains. *Proceedings of the Institution of Civil Engineers: Water Management*, **158**(2), 81–88.

Lin L.-S. and Blatchley III, E. R. (2001). UV Dose distribution characterization using fractal concepts for system performance evaluation. *Water Science and Technology*, **43**(11), 181–188.

Lin Y. and Zeng G. (2000). Study on optimal design model for city wastewater treatment plant considering the effect of uncertainty. *Hunan Daxue Xuebao/Journal of Hunan University Natural Sciences*, **27**(4), 94–100.

Liu Y. and Hong H. P. (2000). Optimal design for wastewater treatment. 2000 Annual Conference – Canadian Society for Civil Engineering, June 7, 2000–June 10, 2000, London, Ontario, Canada, Canadian Society for Civil Engineering.

Liu H.-H., Zhang H.-Z. and Bai X.-Y. (2008a). A model for the optimal design of a wastewater treatment system for constructed wetlands. *Water Resources Protection*, **24**(3), 35–38.

Liu Y., Guo H., Zhou F. and Qin X. (2008b). Inexact chance-constrained linear programming model for optimal water pollution management at the watershed scale. *Journal of Water Resources Planning and Management*, **134**(4), 347–356.

Lodge B. N. and Judd S. J. (2006). A statistical approach to the optimisation of membrane operation. *Water and Environment Journal*, **20**(2), 96–100.

Magbanua B. S., Jr. (2004). Probabilistic modeling of activated sludge systems: Strategies for reducing process uncertainty. 2004 World Water and Environmental Resources Congress: Critical Transitions in Water and Environmental Resources Management, June 27, 2004–July 1, 2004, Salt Lake City, UT, United States, American Society of Mechanical Engineers.

Magbanua B. S., Jr. (2007). Probabilistic modeling of carbon and nitrogen removal during wastewater treatment using activated sludge model no. 1. 3rd IASTED International Conference on Environmental Modelling and Simulation, EMS 2007, August 20, 2007–August 22, 2007, Honolulu, HI, United States, Acta Press.

Makinia J. and Wells S. A. (2000). A general model of the activated sludge reactor with dispersive flow – II. Model verification and application. *Water Research*, **34**(16), 3997–4006.

Martin-Sistac C., Escudero G. and Graells M. (2007). General framework for solving the design and operation of wastewater treatment networks. *Computer Aided Chemical Engineering*, **24**, 1325–1330.

Moles C. G., Gutierrez G., Alonso A. A. and Banga J. R. (2003). Integrated process design and control via global optimization: a wastewater treatment plant case study. *Chemical Engineering Research and Design*, **81**(5), 507–517.

Morris C. D. and Moore G. T. (2003). Preparing for CMOM; A New approach using Stochastic RDII modeling. World Water and Environmental Resources Congress 2003, June 23, 2003–June 26, 2003, Philadelphia, PA, United States, American Society of Civil Engineers.

Munoz I., Rodriguez A., Rosal R. and Fernandez-Alba A. R. (2009). Life cycle assessment of urban wastewater reuse with ozonation as tertiary treatment. A focus on toxicity-related impacts. *Science of the Total Environment*, **407**(4), 1245–1256.

Neumann M. (2007). Uncertainty Analysis for Performance Evaluation and Design of Urban Water Infrastructure. ETH Zurich, Switzerland.

Neumann M. B., Gujer W. and von Gunten U. (2009). Global sensitivity analysis for model-based prediction of oxidative micropollutant transformation during drinking water treatment. *Water Research*, **43**(4), 997–1004.

Peters G. M. and Rowley H. V. (2009). Environmental comparison of biosolids management systems using life cycle assessment. *Environmental Science and Technology*, **43**(8), 2674–2679.

Poplewski G. and Jezowski J. (2004). Water usage minimisation in existing water networks – The use of stochastic optimisation. *Inzynieria Chemiczna I Procesowa*, **25**(3 III), 1545–1550.

Poplewski G. and Jezowski J. (2007). A method for designing optimal wastewater treatment network with stochastic optimization. *Inzynieria Chemiczna I Procesowa*, **28**(4), 941–949.

Portielje R., Hvitved-Jacobsen T. and Schaarup-Jensen K. (2000). Risk analysis using stochastic reliability methods applied to two cases of deterministic water quality models. *Water Research*, **34**(1), 153–170.

Poznyak T., Chairez I. and Poznyak A. (2005). Application of a neural observer to phenols ozonation in water: simulation and kinetic parameters identification. *Water Research*, **39**(12), 2611–2620.

Ramirez I. and Steyer J. P. (2008). Modeling microbial diversity in anaerobic digestion. *Water Science and Technology*, **57**(2), 265–270.

Rigopoulos S. and Linke P. (2002). Systematic development of optimal activated sludge process designs. *Computers and Chemical Engineering*, **26**(4–5), 585–597.

Rivas A., Irizar I. and Ayesa E. (2008). Model-based optimisation of wastewater treatment plants design. *Environmental Modelling and Software*, **23**(4), 435–450.

Rousseau D., Verdonck F., Moerman O., Carrrette R., Thoeye C., Meirlaen J. and Vanrolleghem P. A. (2001). Development of a risk assessment based technique for design/retrofitting of WWTPs. *Water Science and Technology*, **43**(7), 287–294.

Rousseau D. P. L., Vanrolleghem P. A. and De Pauw N. (2004). Model-based design of horizontal subsurface flow constructed treatment wetlands: a review. *Water Research*, **38**(6), 1484–1493.

Sadiq R., Husain T., Al-Zahrani A. M., Sheikh A. K. and Farooq S. (2003). Secondary effluent treatment by slow sand filters: performance and risk analysis. *Water, Air, and Soil Pollution*, **143**(1–4), 41–63.

Sadiq R., Al-Zahrani M. A., Sheikh AK. ., Husain T. and Farooq S. (2004). Performance evaluation of slow sand filters using fuzzy rule-based modelling. *Environmental Modelling and Software*, **19**(5), 507–515.

Simonovic S. P. and Verma R. (2008). A new methodology for water resources multicriteria decision making under uncertainty. *Physics and Chemistry of the Earth*, **33**(5), 322–329.

Sin G., De Pauw D. J. W., Weijers S. and Vanrolleghem P. A. (2008). An efficient approach to automate the manual trial and error calibration of activated sludge models. *Biotechnology and Bioengineering*, **100**(3), 516–528.

Sin G., Gernaey K. V., Neumann M. B., van Loosdrecht M. C. M. and Gujer W. (2011). Global sensitivity analysis in WWTP model applications – prioritising sources of uncertainty. *Water Research*, **45**, 639–651.

Tansel B. (1999). Propagation of parameter uncertainties to system dependability. *Civil Engineering and Environmental Systems*, **16**(1), 19–35.

Tian K.-J., Liu X.-A., Kennedy M. D., Schippers J. C. and Vanrolleghem P. A. (2004). Modelling the biological performance of a side-stream membrane bioreactor using ASM1. *Journal of Environmental Sciences*, **16**(6), 1020–1023.

Tsai J. C. C., Chen V. C. P., Beck M. B. and Chen J. (2004). Stochastic dynamic programming formulation for a wastewater treatment decision-making framework. *Annals of Operations Research*, **132**(1–4), 207–221.

Vanrolleghem P., Insel G., Petersen B., Sin G., Nopens I., Dovermann H., Weijers S. and Gernaey K. (2003). A comprehensive model calibration procedure for activated sludge models. Proceedings of WEFTEC, Alexandria, VA, USA.

Vanrolleghem P. and Van Daele M. (1994). Optimal experimental design for structure characterization of biodegradation models: On-line implementation in a respirographic biosensor. Proceedings of the 17th Biennial Conference of the International Association on Water Quality. Part 1, July 24, 1994–July 30, 1994, Budapest, Hung, Pergamon Press Inc.

Vanrolleghem P. A., Jeppsson U., Carstense J., Carlssont B. and Olsson G. (1996). Integration of wastewater treatment plant design and operation – A systematic approach using cost functions. Proceedings of the 1996 18th Biennial Conference of the International Association on Water Quality. Part 2, June 23, 1996–June 28, 1996, Singapore, Singapore, Pergamon Press Inc.

Vanrolleghem P. A., Spanjers H., Petersen B., Ginsestet P. and Takacs I. (1999). Estimating (combinations of) activated sludge model no. 1 parameters and components by respirometry. *Water Science and Technology*, **39** (1), 195–214.

Vaughan W. J., Russell C. S., Rodriguez D. J. and Darling A. C. (2000). Uncertainty in cost-benefit analysis based on referendum contingent valuation. *Impact Assessment and Project Appraisal*, **18**(2), 125–137.

Volcke E. I. P., Loccufier M., Vanrolleghem PA. . and Noldus E. J. L. (2006). Existence, uniqueness and stability of the equilibrium points of a SHARON bioreactor model. *Journal of Process Control*, **16**(10), 1003–1012.

Vollertsen J., Hvitved-Jacobsen T. and Nielsen A. H. (2005a). Stochastic modeling of chemical oxygen demand transformations in gravity sewers. *Water Environment Research*, **77**(4), 331–339.

Vollertsen J., Nielsen A. H., Yan W. and Hvitved-Jacobsen T. (2005b). Effects of in-sewer processes: A stochastic model approach. *Water Science and Technology*, **52**(3), 171–180.

Von Sperling M. (1993). Parameter estimation and sensitivity analysis of an activated sludge model using Monte Carlo simulation and the analyst's involvement. Proceedings of the 6th JAWO Workshop on Instrumentation, Control and Automation of Water and Wastewater Treatment and Transport Systems, June 17, 1993–June 25, 1993, Banff, Canada, Pergamon Press Inc.

Von Sperling M. (1994). Calibration of poorly identifiable systems: application to activated sludge model. *Journal of Environmental Engineering*, **120**(3), 625–644.

Von Sperling M. (1996). Design of facultative ponds based on uncertainty analysis. Proceedings of the 1995 3rd International Symposium on Waste Stabilization Ponds: Technology and Applications, March 27, 1995–March 31, 1995, Water Science and Technology, **33**(7), 41–47.

Von Sperling M. (1999). A critical analysis of classical design equations for waste stabilization lagoons and other wastewater treatment systems. *Water Environment Research*, **71**(6), 1240–1243.

Von Sperling M. (2003). Influence of the dispersion number on the estimation of coliform removal in ponds. *Water Science and Technology*, **48**(2), 181–188.

Wang J., Guo H., Liu L., Hao M., Zhang M., Lu X. and Xing K.(2004). Inexact multi-objective programming approach for strategic environmental assessment on regional development plan. *Progress in Natural Science*, **14**(11), 950–959.

Wang L.-L., Zeng G.-M., Huang G.-H., Yang F. and Wu L.(2006). Selection of technology for sewage treatment considering the effect of uncertainty. *Hunan Daxue Xuebao/Journal of Hunan University Natural Sciences*, **33** (6), 115–118.

Wang L., Chen J., Zeng S. and Du J. (2009). Numerical modeling-based municipal wastewater treatment process screening methodology. *Qinghua Daxue Xuebao/Journal of Tsinghua University*, **49**(3), 377–380.

Wen C.-G. and Lee C.-S. (1999). Development of a cost function for wastewater treatment systems with fuzzy regression. *Fuzzy Sets and Systems*, **106**(2), 143–153.

Werner T. and Kadlec R. (2000). Stochastic simulation of partially-mixed, event-driven treatment wetlands. *Ecological Engineering*, **14**(3), 253–267.

Willems P. (2008). Quantification and relative comparison of different types of uncertainties in sewer water quality modeling. *Water Research*, **42**(13), 3539–3551.

Wu E. M.-Y. (2002). Stochastic integer programming analysis for wastewater treatment plant design. Third International Conference on Management Information Systems Incorporating GIS and Remote Sensing, MIS 2002, April 24, 2002–April 26, 2002, Halkidiki, Greece, WIT Press.

Yetilmezsoy K. and Sapci-Zengin Z. (2009). Stochastic modeling applications for the prediction of COD removal efficiency of UASB reactors treating diluted real cotton textile wastewater. *Stochastic Environmental Research and Risk Assessment*, **23**(1), 13–26.

Zeng G.-M., Lin Y.-P., Qin X., Huang G.-H., Li J.-B. and Jiang R. (2004). Optimum municipal wastewater treatment plant design with consideration of uncertainty. *Journal of Environmental Sciences*, **16**(1), 126–131.

Zhou F., Chen G., Guo H. and Liu Y. (2008). Modified interval linear programming for lake watershed management. *Huanjing Kexue Xuebao / Acta Scientiae Circumstantiae*, **28**(8), 1688–1698.

C.5 UNCERTAINTY IN WASTEWATER TREATMENT PLANT OPERATIONAL CONTROL DATA AND METHODS OF ADDRESSING IN ONLINE CONTROL

Aguilar-Lopez R. (2008). Robust generic model control for dissolved oxygen in activated sludge wastewater plant. *Chemical and Biochemical Engineering Quarterly*, **22**(1), 71–79.

Åkrjanc I. (2009). Confidence interval of fuzzy models: an example using a waste-water treatment plant. *Chemometrics and Intelligent Laboratory Systems*, **96**(2), 182–187.

Aksakalli V. and Ursu D. (2006). Control of nonlinear stochastic systems: Model-free controllers versus linear quadratic regulators. 45th IEEE Conference on Decision and Control 2006, CDC, December 13, 2006–December 15, 2006, San Diego, CA, United States, Institute of Electrical and Electronics Engineers Inc.

Alcaraz-Gonzalez V. and lvarez V. (2007). Robust Nonlinear Observers for Bioprocesses: Application to Wastewater Treatment. Department of Chemical Engineering, University of Guadalajara, Guadalajara, Jalisco, Mexico.

Alcaraz-Gonzalez V., Maloum A., Haramnd J., Rapaport A., Steyer J. P., Gonzalez-Alvarez V. and Pelayo-Ortiz C. (2000). Robust interval-based siso and simo regulation for a class of highly uncertain bioreactors: Application to the anaerobic digestion. 39th IEEE Confernce on Decision and Control, December 12, 2000–December 15, 2000, Sydney, NSW, Australia, Institute of Electrical and Electronics Engineers Inc.

Alcaraz-Gonzalez V., Harmand J., Rapaport A., Steyer J. P., Gonzalex-Alvarez V. and Pelayo-Ortiz C. (2002). Software sensors for highly uncertain WWTPs: a new approach based on interval observers. *Water Research*, **36**(10), 2515–2524.

Alcaraz-Gonzalez V., Harmand J., Rapaport A., Steyer J. P., Gonzalex-Alvarez V. and Pelayo-Ortiz C. (2005a). Application of a robust interval observer to an anaerobic digestion process. *Developments in Chemical Engineering and Mineral Processing*, **13**(3–4), 267–278.

Alcaraz-Gonzalez V., Harmand J., Rapaport A., Steyer J. P., Gonzalex-Alvarez V. and Pelayo-Ortiz C. (2005b). Robust interval-based regulation for anaerobic digestion processes. *Water Science and Technology*, **52**(1–2), 449–456.

Alcaraz-Gonzalez V., Salazar-Pena R., Gouze J.-L. and Steyer J. P. (2005c). A tunable multivariable nonlinear robust observer for biological systems. *Comptes Rendus Biologies*, **328**(4), 317–325.

Antonelli R. and Astolfi A. (2000). Nonlinear controllers design for robust stabilization of continuous biological reactors. Proceedings of the 2000 IEEE International Conference on Control Applications, Anchorage, USA, IEEE.

Antonelli R., Harmand J., Steyere J. P. and Astolfic A. (2002). Output feedback regulation of an anaerobic digestion process. 2002 American Control Conference, May 8, 2002–May 10, 2002, Anchorage, AK, United States, Institute of Electrical and Electronics Engineers Inc.

Antonelli R., Harmand J., Steyer J. P. and Astolfi A. (2003). Set-point regulation of an anaerobic digestion process with bounded output feedback. *IEEE Transactions on Control Systems Technology*, **11**(4), 495–504.

Aschemann H., Rauh A., Kletting M. and Hofer E. P. (2007). Flatness-based control of a simplified wastewater treatment plant. Joint 2006 IEEE Conference on Control Applications (CCA), Computer-Aided Control Systems Design Symposium (CACSD) and International Symposium on Intelligent Control (ISIC), October 4, 2006–October 6, 2006, Munich, Germany, Institute of Electrical and Electronics Engineers Inc.

Astorga Zaragoza C. M., Othman S. and Hammouri H., (2000). Moving horizon control of a denitrification reactor, in: WSES/International Conferences AMTA 2000-MCBC 2000-MCBE 2000, Jamaica, pp. 931–936.

Ben Yousse C. and Torres Zuniga J. I. (2002). Stable model-reference adaptive control and estimation of a wastewater treatment benchmark. Proceedings of the IEEE International Conference on Systems, Man and Cybernetics, 6–9 October 2002, Yasmine Hammamet, Tunisia.

Benedetti L., De Baets B., Nopens I. and Vanrolleghem P. A. (2010). Multi-criteria analysis of wastewater treatment plant design and control scenarios with the Benchmark Simulation Model No. 2 under uncertainty. *Environmental Modelling & Software*, **25**(5), 616–621.

Betancur M. J., Moreno J. A., Moreno-Andrade I. and Buitron G. (2006). Practical optimal control of fed-batch bioreactors for the waste water treatment. *International Journal of Robust and Nonlinear Control*, **16**(3), 173–190.

Chachuat B. and Bernard O. (2006). Probabilistic observers for a class of uncertain biological processes. *International Journal of Robust and Nonlinear Control*, **16**(3), 157–171.

Chachuat B., Roche N. and Latifi R. (2003). Reduction of the ASM 1 model for optimal control of small-size activated sludge treatment plants Reduction du modele ASM 1 pour la commande optimale des petites stations d'epuration a boues activees. *Revue des Sciences de L'Eau*, **16**(1), 5–26.

Chang N.-B., Chen W. C. and Shieh W. K. (2001). Optimal control of wastewater treatment plants via integrated neural network and genetic algorithms. *Civil Engineering and Environmental Systems*, **18**(1), 1–17.

Chen W. C., Chang N.-B. and Shieh W. K. (2001). Advanced hybrid fuzzy-neural controller for industrial wastewater treatment. *Journal of Environmental Engineering*, **127**(11), 1048–1059.

Chong H. and Walley W. (1996). Rule-based versus probabilistic approaches to the diagnosis of faults in wastewater treatment processes. *Artificial Intelligence in Engineering*, **10**(3), 265–273.

Chui J. and Li J. Y. (2000). Long-term rainfall generation for real time control planning of combined sewer systems. 2000 Annual Conference – Canadian Society for Civil Engineering, June 7, 2000–June 10, 2000, London, Ontario, Canada, Canadian Society for Civil Engineering.

Comas J., Dzeroski S., Gibert K., Rodriguez-Roda I. and Sanchez-Marre M. (2001). Knowledge discovery by means of inductive methods in wastewater treatment plant data. *AI Communications*, **14**(1), 45–62.

Comas J., Rodriguez-Roda I., Poch M., Gernaey K. V., Rosen C. and Jeppsson U. (2006). Demonstration of a tool for automatic learning and re-use of knowledge in the activated sludge process. *Water Science and Technology*, **53**(4–5), 303–311.

Cong Q.-M., Chai T.-Y. and Yu W. (2009). Modeling wastewater treatment plant via hierarchical neural networks. *Kongzhi Lilun Yu Yinyong/Control Theory and Applications*, **26**(1), 8–14.

Cox C., Fletcher I. and Adgar A. (2001). ANN-based sensing and control developments in the water industry: A decade of innovation. Proceedings of the 2001 IEEE International Symposium on Intelligent Control ISIC '01, September 5, 2001–September 7, 2001, Mexico City, Mexico, Institute of Electrical and Electronics Engineers Inc.

Dimitrova N. and Krastanov M. (2006). Nonlinear adaptive control of an uncertain wastewater treatment model. 12th GAMM – IMACS International Symposium on Scientific Computing, Computer Arithmetic and Validated Numerics, SCAN 2006, Conference Post-Proceedings.

El-Din A. G. and Smith D. W. (2000). Real-time waste water flow prediction using an ANN. 2000 Annual Conference Abstracts – Canadian Society for Civil Engineering.

El-Din A. G. and Smith D. W. (2002). A neural network model to predict the wastewater inflow incorporating rainfall events. *Water Research*, **36**(5), 1115–1126.

Ferrer A., Aguado D., Vidal-Puig S. and Montalban J. P. (2008). PLS: A versatile tool for industrial process improvement and optimization. *Applied Stochastic Models in Business and Industry*, **24**(6), 551–567.

Fibrianto H., Mazouni D., Ignatova M., Herveau M., Harmand J. and Dochain D. (2008). Dynamical modelling, identification and software sensors for SBRs. *Mathematical and Computer Modelling of Dynamical Systems*, **14** (1), 17–26.

Fiter M., Güell D., Comas J., Colprim J., Poch M. and Rodriguez-Roda I. (2005). Energy saving in a wastewater treatment process: an application of fuzzy logic control. *Environmental Technology*, **26**(11), 1263–1270.

Flores X., Comas J., Rodriguez-Roda I., Jimenez Esteller L. and Gernaey K. V. (2007). Application of multivariable statistical techniques in plant-wide WWTP control strategies analysis. *Water Science and Technology*, **56**(6), 75–83.

Flores-Alsina X., Rodriguez-Roda I., Sin G. and Gernaey K. V. (2009). Uncertainty and sensitivity analysis of control strategies using the benchmark simulation model No1 (BSM1). *Water Science Technology*, **59**(3), 491–499.

Flores-Tlacuahuac A., Esparza M. and Lopez-Negrete de la Fuente R. (2009). Bifurcation behavior of a large scale waste water treatment plant. *Industrial & Engineering Chemistry Research*, **48**(5), 2605–2615.

Garcia-Sanz M., Eguinoa I., Gil-Martinez M., Irizar I. and Ayesa E. (2008). MIMO quantitative robust control of a wastewater treatment plant for biological removal of nitrogen and phosphorus. In proceedings of the 16th Mediterranean Conference on Control and Automation, 2008: MED '08; 25–27 June 2008, Ajaccio, Corsica, France.

Giona M. and Paladino O. (1994). Bifurcation-Analysis and stability of a controlled CSTR. *Computers & Chemical Engineering*, **18**(9), 877–887.

Gomez-Quintero C. S. and Queinnec I. (2002). Robust estimation for an uncertain linear model of an activated sludge process. Proceedings of the 2002 IEEE International Conference on Control Applications, September 18–20, 2002, Glasgow, United Kingdom, Institute of Electrical and Electronics Engineers Inc.

Gomez-Quintero C.-S. and Queinnec I. (2002b). Robust estimation for an uncertain linear model of an activated sludge process. Proceedings of the 2002 IEEE International Conference on Control Applications, September 18, 2002–September 20, 2002, Glasgow, United Kingdom, Institute of Electrical and Electronics Engineers Inc.

Guergachi A. A. and Patry G. G. (2006). Constructing a model hierarchy with background knowledge for structural risk minimization: application to biological treatment of wastewater. *IEEE Transactions on Systems. Man, and Cybernetics Part A:Systems and Humans*, **36**(2), 373–383.

Hadj-Sadok M. Z. and Gouze J.-L. (1998). Bounds estimations for uncertain models of wastewater treatment. Proceedings of the 1998 IEEE International Conference on Control Applocations. Part 1 (of 2), September 1, 1998–September 4, 1998, Trieste, Italy, IEEE.

Hadj-Sadok M. Z. and Gouze J. L. (2001). Estimation of uncertain models of activated sludge processes with interval observers. *Journal of Process Control*, **11**(3), 299–310.

Holenda B., Domokos E., Redey A. and Fazakas J. (2007). Aeration optimization of a wastewater treatment plant using genetic algorithm. *Optimal Control Applications and Methods*, **28**(3), 191–208.

Hu Y.-L. and Qiao J.-F. (2004). Fuzzy neural network control of uncertain parameters system. Proceedings of the World Congress on Intelligent Control and Automation 15–19 June 2004, Hangzhou, China.

Johnson M. A. and Sanchez A. (2003). Process control loop tuning and monitoring using LQG optimality with applications in wastewater treatment plant. Fourth International Conference on Control and Automation, June 10, 2003–June 12, 2003, Montreal, Québec, Canada, Institute of Electrical and Electronics Engineers Inc.

Karuppiah R., Grossmann I. E. (2008). Global optimization of multiscenario mixed integer nonlinear programming models arising in the synthesis of integrated water networks under uncertainty. Process Systems Engineering: Contributions on the State-of-the-Art Selected extended Papers from ESCAPE '16/PSE 2006. *Computers and Chemical Engineering*, **32**(1–2), 145–160.

Kim Y., Bae H., Poo K., Moon T., Kim S. and Kim C. (2006). Soft sensor using PNN model and rule base for wastewater treatment plant. 3rd International Symposium on Neural Networks, ISNN 2006 – Advances in Neural Networks, May 28, 2006–June 1, 2006, Chengdu, China, Springer-Verlag.

Kolodyazhniy V., Klawonn F. and Tschumitschew K. (2007). A neuro-fuzzy model for dimensionality reduction and its application. *International Journal of Uncertainty, Fuzziness and Knowlege-Based Systems*, **15**(5), 571–593.

Lamrini B., Benhammou A., Le Le Lann M.-V. and Karama A. (2005). A neural software sensor for online prediction of coagulant dosage in a drinking water treatment plant *Transactions of the Institute of Measurement and Control*, **27**(3), 195–213.

Lardon L. and Steyer J. P. (2007). Advances in Diagnosis of Biological Anaerobic Wastewater Treatment Plants. Selected Topics in Dynamics and Control of Chemical and Biological Processes, Springer, Berlin, Heidelberg.

Lardon L., Punal A. and Steyer J. P. (2004). On-line diagnosis and uncertainty management using evidence theory – experimental illustration to anaerobic digestion processes. *Dynamics, Monitoring Control and Optimization of Biological Journal of Process Control*, **14**(7), 747–763.

Lardon L., Punal A., Martinez J. A. and Steyer J. P. (2005). Modular expert system for the diagnosis of operating conditions of industrial anaerobic digestion plants. *Water Science and Technology*, **52**(1–2), 427–433.

Lee D. S. and Vanrolleghem P. A. (2003). Monitoring of a sequencing batch reactor using adaptive multiblock principal component analysis. *Biotechnology and Bioengineering*, **82**(4), 489–497.

Lee D. S. and Vanrolleghem P. A. (2004). Adaptive consensus principal component analysis for on-line batch process monitoring. *Advanced Environmental Monitoring and Assessment*, **92**(1–3), 119–135.

Lee J.-M., Yoo C., Choi S. W., Vanrolleghem P. A. and Lee I.-B. (2004a). Nonlinear process monitoring using kernel principal component analysis. *Chemical Engineering Science*, **59**(1), 223–234.

Lee T. T., Wang F. Y. and Newell R. B. (2004b). Robust multivariable control of complex biological processes. *Journal of Process Control*, **14**(2), 193–209.

Lee D. S., Park J. M. and Vanrolleghem P. A. (2005). Adaptive multiscale principal component analysis for on-line monitoring of a sequencing batch reactor. *Journal of Biotechnology*, **116**(2), 195–210.

Lennox J. and Rosen C. (2002). Adaptive multiscale principal components analysis for online monitoring of wastewater treatment. *Water Science & Technology*, **45**(4), 227–235.

Marsili-Libelli S. (2006). Control of SBR switching by fuzzy pattern recognition. *Water Research*, **40**(5), 1095–1107.

Marsili-Libelli S. and Giunti L. (2002). Fuzzy predictive control for nitrogen removal in biological wastewater treatment. *Water Science and Technology*, **45**(4–5), 37–44.

Marsili-Libelli S., Spagni A. and Susini R. (2008). Intelligent monitoring system for long-term control of sequencing batch reactors. *Water Science and Technology*, **57**(3), 431–438.

Meirlaen J. and Vanrolleghem P. A. (2002). Model reduction through boundary relocation to facilitate real-time control optimisation in the integrated urban wastewater system. *Water Science and Technology*, **45**(4–5), 373–381.

Meirlaen J., Van Assel J. and Vanrolleghem P. A. (2002). Real time control of the integrated urban wastewater system using simultaneously simulating surrogate models. *Water Science and Technology*, **45**(3), 109–116.

Mingzhi H., Jinquan W., Yongwen M., Weijiang L. and Xiaofei S. (2009). Control rules of aeration in a submerged biofilm wastewater treatment process using fuzzy neural networks. *Expert Systems with Applications*, **36**(7), 10428–10437.

Moreno J. (1999). Optimal time control of bioreactors for the wastewater treatment. *Optimal Control Applications and Methods*, **20**(3), 145–164.

Muller A., Marsili-Libelli S., Aivasidis A., Lloyd T., Kroner S. and Wandrey C. (1997). Fuzzy control of disturbances in a wastewater treatment process. *Water Research*, **31**(12), 3157–3167.

Nejjari F., Roux G., Dahhou B. and Benhammou A. (1999). Estimation and optimal control design of a biological wastewater treatment process. *Mathematics and Computers in Simulation*, **48**(3), 269–280.

Paladino O. (1999). A fuzzy operating database to support diagnosis on a wastewater treatment plant. *Chemical and Biochemical Engineering Quarterly*, **13**(1), 1–8.

Pan Y., Yoo C., Lee J. H. and Lee I.-B. (2002). Process monitoring of continuous processes with periodic operation patterns. 2002 American Control Conference, May 8, 2002–May 10, 2002, Anchorage, AK, United States, Institute of Electrical and Electronics Engineers Inc.

Pan Y., Yoo C., Lee J. H. and Lee I.-B. (2004). Process monitoring for continuous process with periodic characteristics. *Journal of Chemometrics*, **18**(2), 69–75.

Rauh A., Kletting M., Aschemann H. and Hofer E. P. (2005). Robust controller design for bounded state and control variables and uncertain parameters using interval methods. Proceedings of the 5th International Conference on Control and Automation, 26*–29 June 2005, Budapest, Hungary.

Riesco J., Calvo J. and Martin-Sanchez J. M. (2002). Adaptive predictive expert (ADEX) control: Application to waste water treatment plants. IEEE Conference on Control Applications – Proceedings, Glasgow, UK.

Riesco J., Calvo J. and Martin-Sanchez J. M. (2003). ADEX optimization system for the biological phase of a water treatment plant. Technical Papers of ISA: Safety Instrumented Systems for the Process Industry.

Rosen C. and Lennox J. (2001). Multivariate and multiscale monitoring of wastewater treatment operation. *Water Research*, **35**(14), 3402–3410.

Rosen C., Rieger L., Jeppsson U. and Vanrolleghem P. A. (2008). Adding realism to simulated sensors and actuators. *Water Science and Technology*, **57**(3), 337–344.

Sahely B. S. G. E. and Bagley D. M. (2001). Diagnosing upsets in anaerobic wastewater treatment using Bayesian belief networks. *Journal of Environmental Engineering*, **127**(4), 302–310.

Sangüesa R. and Burrell P. (2000). Application of Bayesian network learning methods to waste water treatment plants. *Applied Intelligence*, **13**(1), 19–40.

Sotomayor O. A. Z., Park S. W. and Garcia C. (2002). Software sensor for on-line estimation of the microbial activity in activated sludge systems. *ISA Transactions*, **41**(2), 127–143.

Steyer J.-P., Lardon L. and Bernard O. (2004). Sensors network diagnosis in anaerobic digestion processes using evidence theory. *Water Science and Technology*, **50**(11), 21–29.

Tenno R. and Uronen P. (1996). Stochastic control for wastewater treatment processes. *Control Engineering Practice*, **4**(5), 601–614.

Thomann M., Rieger L., Frommhold S. and Siegrist H. (2002). An efficient monitoring concept with control charts for on-line sensors. *Water Science and Technology*, **46**(4–5), 107–116.

Tomita R., Park S. and Sotomayor O. A. Z. (2002). Analysis of activated sludge process using multivariate statistical tools—a PCA approach. *Chemical Engineering Journal*, **90**(3), 283–290.

Wik T., Fransson C.-M. and Lennartsson B. (2003). Feedforward Feedback Controller Design for Uncertain Systems. 42nd IEEE Conference on Decision and Control, December 9, 2003–December 12, 2003, Maui, HI, United States, Institute of Electrical and Electronics Engineers Inc.

Xie S.-G. and Zhou L.-F. (2008). Fuzzy adaptive generalized predictive control for wastewater treatment process. *Huadong Ligong Daxue Xuebao / Journal of East China University of Science and Technology*, **34**(6), 854–858.

Yiannakopoulou T. V. and Kaimakamidou V. (2009). Testing the reliability of protozoa as indicators of wastewater treatment plant performance. *Fresenius Environmental Bulletin*, **18**(2), 146–157.

Yoo C. K., Vanrolleghem P. A. and Lee I.-B. (2003). Nonlinear modeling and adaptive monitoring with fuzzy and multivariate statistical methods in biological wastewater treatment plants. *Journal of Biotechnology*, **105**(1–2), 135–163.

Yoo C. K., Lee D. S. and Vanrolleghem P. A. (2004). Application of multiway ICA for on-line process monitoring of a sequencing batch reactor. *Water Research*, **38**(7), 1715–1732.

Yoo C. K., Villez K., Lee I.-B. and Vanrolleghem P. A. (2006). Multivariate nonlinear statistical process control of a sequencing batch reactor. *Journal of Chemical Engineering of Japan*, **39**(1), 43–51.

Yoo C. K., Villez K., Lee I.-B., Rosen C. and Vanrolleghem P. A. (2007). Multi-model statistical process monitoring and diagnosis of a sequencing batch reactor. *Biotechnology and Bioengineering*, **96**(4), 687–701.

Yordanova S., Petrova R., Mastorakis N. E. and Mladenov V. (2006). Sugeno predictive neuro-fuzzy controller for control of nonlinear plant under uncertainties. *WSEAS Transactions on Systems*, **5**(8), 1814–1821.

Zacharof A. I., Schutze M. and Butler D. (2003). A methodology for assessing the operational potential of the urban wastewater system using integrated modelling. *Water Supply*, **3**(1–2), 271–277.

Zhu G.-B. and Peng Y.-Z. (2004). Application and development of intelligent control technology in urban wastewater treatment system. *Harbin Gongye Daxue Xuebao/Journal of Harbin Institute of Technology*, **36**(3), 328–330+374.

C.6 UNCERTAINTY IN THE FATE OF POLLUTANTS IN THE ENVIRONMENT AND RESULTING IN REGULATORY (WWTP EFFLUENT STANDARDS) ISSUES

Abdulla F. A. and Al-Ghazzawi Z. D. (1998). Assessment of methane emissions from industrial and municipal wastewater treatment plants in Jordan. *Journal of the Institution of Engineers (India): Environmental Engineering Division*, **79**(1), 1–3.

Abdulla F. A. and Al-Ghazzawi Z. D. (2000). Methane emissions from domestic waste management facilities in Jordan – applicability of IPCC methodology. *Journal of the Air and Waste Management Association*, **50**(2), 234–239.

Ahlman S. (2006). Modelling of substance flows in urban drainage systems. *Doktorsavhandlingar vid Chalmers Tekniska Hogskola*, **2441**, 1–90.

Alhumoud J. M. and Al-Sharifi F. A. (2002). Probability analysis of Kuwait's wastewater treatment plant's influent and effluent. *Fluid – Particle Separation Journal*, **14**(2), 95–104.

An Y.-J., Yoon C. G., Jung K.-W. and Ham J.-H. (2007). Estimating the microbial risk of *E. coli* in reclaimed wastewater irrigation on paddy field. *Environmental Monitoring and Assessment*, **129**(1–3), 53–60.

Andreadakis A., Mamais D. and Gavalakis E. (2006). Evaluation of treatment schemes appropriate for wastewater reuse in Greece. *International Journal of Environmental Technology and Management*, **6**(5), 423–433.

Barnett V. and O'Hagan A. (1997). Setting Environmental Standards: The Statistical Approach to Handling Uncertainty and Variation (1st ed.). Chapman and Hall/CRC. https://doi.org/10.1201/9780367812546

Baxter C. W., Lence B. J. and Coffey B. M. (2003). Analysing operational risk in potable water supply using conditional reliability. In Proceedings of the World Water and Environmental Resources Congress, 23–26 June 2003, Philadelphia, USA.

Benedetti L., Blumensaat F., Bonisch G., Dirckx G., Jardin N., Krebs P. and Vanrolleghem P. A. (2005). Systems analysis of urban wastewater systems – Two systematic approaches to analyse a complex system. *Water Science and Technology*, **52**(12), 171–179.

Benedetti L., Dirckx G., Bixio D., Thoeye C. and Vanrolleghem P. A. (2006). Substance flow analysis of the wastewater collection and treatment systems. *Urban Water Journal*, **3**(1), 33–42.

Benedetti L., Meirlaen J., Facchi A., Gandolfi C. and Vanrolleghem P. A. (2007). Dynamic integrated water quality modelling: a case study of the Lambro River, northern Italy. *Water SA*, **33**(5), 627–632.

Benedetti L., Dirckx G., Bixio D., Thoeye C. and Vanrolleghem P. A. (2008). Environmental and economic performance assessment of the integrated urban wastewater system. *Journal of Environmental Management*, **88**, 1262–1272.

Benedetti L., Prat P., Nopens I., Poch M., Turon C., De Baets B. and Comas J. (2009). A new rule generation method to develop a DSS for integrated management at river basin scale. *Water Science and Technology*, **60**(8), 2035–2040.

Benedetti L., De Keyser W., Nopens I. and Vanrolleghem P. A. (2010). Probabilistic modelling and evaluation of waste water treatment plant upgrades in the EU water framework directive context. *Journal of Hydroinformatics*, **12**(4), 380–395.

Benetto E., Nguyen D., Lohmann T., Schmitt B. and Schosseler P. (2009). Life cycle assessment of ecological sanitation system for small-scale wastewater treatment. *Science of the Total Environment*, **407**(5), 1506–1516.

Bischoff W., Huang M.-N. L. and Yang L. (2006). Growth curve models for stochastic modeling and analyzing of natural disinfection of wastewater. *Environmetrics*, **17**(8), 827–847.

Bo Y. (1998). Probabilistic management model for water quality. Proceedings of the 1998 International Water Resources Engineering Conference. Part 2 (of 2), August 3, 1998–August 7, 1998, Memphis, TN, USA, ASCE.

Boeije G., Vanrolleghem P. and Matthies M. (1997). A geo-referenced aquatic exposure prediction methodology for 'down-the-drain' chemicals. Proceedings of the 1997 4th IAWQ International Symposium on Systems Analysis and Computing in Water Quality Management, June 17, 1997–June 20, 1997, Quebec, Canada, Elsevier Science Ltd.

Bradley A. A., Holly F. M., Walker W. K. and Wright S. A. (1998). Estimation of water temperature exceedance probabilities using thermo-hydrodynamic modeling. *Journal of the American Water Resources Association*, **34** (3), 467–480.

Breton R. L., Teed R. S. and Moore D. R. J. (2003). An ecological risk assessment of phenol in the aquatic environment. *Human and Ecological Risk Assessment*, **9**(2), 549–568.

Buzas K. (1999). Nutrient balances of the Danube basin: The role of uncertainties. Proceedings of the 1999 8th IAWQ River Basin Conference, Budapest, Hungary, September 13 – 16, 1998. *Water Science and Technology*, **40**(10), 51–58.

Carrasco I. J. and Chang S. Y. (2005). Random Monte Carlo simulation analysis and risk assessment for ammonia concentrations in wastewater effluent disposal. *Stochastic Environmental Research and Risk Assessment*, **19**(2), 134–145.

Cata C. and Mohrlok U. (2006). A random-walk approach for simulating wastewater transport and transformations in the unsaturated zone. ModelCARE 2005, June 6, 2005–June 9, 2005, Hague, Netherlands, IAHS Press.

Choi K. W., Lee J. H. W., Kwok K. W. H. and Leung K. M. Y. (2009). Integrated stochastic environmental risk assessment of the Harbour Area Treatment Scheme (HATS) in Hong Kong. *Environmental Science and Technology*, **43**(10), 3705–3711.

Chu J., Chen J., Wang C. and Fu P. (2004). Wastewater reuse potential analysis: implications for China's water resources management. *Water Research*, **38**(11), 2746–2756.

Clement A. and Buzas K. (1999). Use of ambient water quality data to refine emission estimates in the Danube basin. Proceedings of the 1999 8th IAWQ River Basin Conference, Budapest, Hungary, September 13, 1998 – September 16, 1998. *Water Science and Technology*, **40**(10), 35–42.

Clement L., Thas O., Vanrolleghem P. A. and Ottoy J. P. (2006). Spatio-temporal statistical models for river monitoring networks. *Water Science and Technology*, **53**(1), 9–15.

Constable M., Charlton M., Jensen F., McDonald K., Craig G. and Taylor K. W. (2003). An ecological risk assessment of ammonia in the aquatic environment. *Human and Ecological Risk Assessment*, **9**(2), 527–548.

Crabtree B., Earp W. and Whalley P. (1996). Demonstration of the benefits of integrated wastewater planning for controlling transient pollution. Proceedings of the 1995 IAWQ Interdisciplinary Symposium on Uncertainty, Risk and Transient Pollution Events, Exeter, UK, July 26–28, 1995. Water Science and Technology, **33**(2), 209–218.

De Azevedo L. G. T., Gates T. K., Fontane D. G., Labadie J. W. and Porto R. L. (2000). Integration of water quantity and quality in strategic river basin planning. *Journal of Water Resources Planning and Management*, **126**(2), 85–97.

De Nooij R. J. W., Lotterman K. M., vand de Sande P. H. J., Pelsma T., Leuven R. S. E. W. and Lenders H. J. R. (2006). Validity and sensitivity of a model for assessment of impacts of river floodplain reconstruction on protected and endangered species. *Environmental Impact Assessment Review*, **26**(8), 677–695.

Dong H.-G., Lin C.-Y. and Chang C.-T. (2008). Simultaneous optimization approach for integrated water-allocation and heat-exchange networks. *Chemical Engineering Science*, **63**(14), 3664–3678.

Donovan E., Unice K., Roberts J. D., Harris M. and Finley B. (2008). Risk of gastrointestinal disease associated with exposure to pathogens in the water of the Lower Passaic River. *Applied and Environmental Microbiology*, **74**(4), 994–1003.

Efroymson R. A., Jones D. S. and Gold A. J. (2007). An ecological risk assessment framework for effects of onsite wastewater treatment systems and other localized sources of nutrients on aquatic ecosystems. *Human and Ecological Risk Assessment*, **13**(3), 574–614.

El-Beshry M. Z. (2008). Sensitivity analysis of simulated dissolved oxygen in bahr hadus drain, Egypt. *Journal of Engineering and Applied Science*, **55**(3), 195–208.

El-Fadel M. and Massoud M. (2001). Methane emissions from wastewater management. *Environmental Pollution*, **114** (2), 177–185.

El-Ghonemy H., Watts L. and Fowler L. (2005). Treatment of uncertainty and developing conceptual models for environmental risk assessments and radioactive waste disposal safety cases. *Environment International*, **31**(1), 89–97.

Elektorowicz M. and Qasaimeh A. (2004). Fuzzy modelling estimation of mercury removal by wetland components. In Proceedings of the Annual Conference of the North American Fuzzy Information Processing Society, 27–30 June 2004, Banff, Canada.

Elektorowicz M., Balanzinski M. and Qasaimeh A. (2002). Application of the artificial intelligence to estimate the constructed wetland response to heavy metal removal. Conference Proceedings – Joint 2002 CSCE/ASCE International Conference on Environmental Engineering – An International Perspective on Environmental Engineering, Canada: Canadian Society for Civil Engineering.

Espinoza F. P. and Enfield C. G. (2005). Stochastic-Lagrangian Model (SLAM) for NAPL remediation. 2005 World Water and Environmental Resources Congress, May 15, 2005–May 19, 2005, Anchorage, AK, United States, American Society of Civil Engineers.

Fan L. and Xu Y. (2008). Interval observer-based monitoring of river water quality. 2007 IEEE International Conference on Control and Automation, ICCA, May 30, 2007–June 1, 2007, Guangzhou, China, Institute of Electrical and Electronics Engineers Inc.

Franks C. A., Mahony M. J. and Franks S. W. (2005). Measuring river health: the uncertainty associated with impact assessment using a single sample biological assessment technique. *IAHS-AISH Publication*, (294), 176–182.

Gelda R. K., Effler S. W. and O'Donnell S. M. (2001). Probabilistic model of ammonia and toxicity status for urban lake. *Journal of Water Resources Planning and Management*, **127**(5), 337–347.

Ghermandi A., Vandenberghe V., Benedetti L., Bauwens W. and Vanrolleghem P. A. (2009). Model-based assessment of the effect of shading on river water quality. *Ecological Engineering*, **35**(1), 92–104.

Gronewold A., Qian S., Wolper R. L. and Reckhow K. H. (2009). Calibrating and validating bacterial water quality models: A Bayesian approach. *Water Research*, **43**(10), 2688–2698.

Hauger M., Rauch W., Linde J. and Mikkelsen P. S. (2002). Model based risk assessment of CSO induced DO depletion in urban rivers. Proceedings of the Ninth International Conference on: Urban Drainage, September 8, 2002–September 13, 2002, Portland, OR, United States, American Society of Civil Engineers.

Jardine C. G. and Hrudey S. E. (1999). Characterizing response variability and method uncertainty in odour perception research. Proceedings of the 1997 5th International Symposium on Off-Flavours in the Aquatic Environment, October 13, 1997–October 16, 1997, Paris, France, Elsevier Ltd.

Johnson A., Purvis Thurow A. and Vietor D. M. (1998). Dairy manure management: an application of probabilistic risk assessment. *Journal of Environmental Quality*, **27**(3), 481–487.

Kapo K. E. and Burton G. A. (2006). A geographic information systems-based, weights-of-evidence approach for diagnosing aquatic ecosystem impairment. *Environmental Toxicology and Chemistry*, **25**(8), 2237–2249.

Li S. and Morioka T. (1999). Optimal allocation of waste loads in a river with probabilistic tributary flow under transverse mixing. *Water Environment Research*, **71**(2), 156–162.

Logan L. and Ponnambalam K. (2001). Integrating risk and performance reliability for flow augmentation. 2001 IEEE International Conference on Systems, Man and Cybernetics, October 7, 2001–October 10, 2001, Tucson, AZ, United States, Institute of Electrical and Electronics Engineers Inc.

Mannina G., Freni G., Vivani G., Saegrov S. and Hafskjold L. S. (2006). Integrated urban water modelling with uncertainty analysis. *Water Science & Technology*, **54**(6–7), 379–386.

Maria G. and Maria C. (2006). Evaluation of risk zones over a river pathway, downstream a release point, under seasonal pollutant biodegradability. *Chemical and Biochemical Engineering Quarterly*, **20**(3), 333–342.

Masliev I. and Somlyody L. (1994). Probabilistic methods for uncertainty analysis and parameter-estimation for dissolved-oxygen models. *Water Science and Technology*, **30**(2), 99–108.

Matthies M., Berlekamp J., Koormann F. and Wagner J.-O. (2001). Georeferenced regional simulation and aquatic exposure assessment.

McFarland A. M. S. and Hauck L. M. (2001). Determining nutrient export coefficients and source loading uncertainty using in-stream monitoring data. *Journal of the American Water Resources Association*, **37**(1), 223–236.

Meirlaen J., Huyghebaert B., Sforzi F., Benedetti L. and Vanrolleghem P. A. (2001). Fast, simultaneous simulation of the integrated urban wastewater system using mechanistic surrogate models. *Water Science and Technology*, **43**(7), 301–309.

Meirlaen J., Van Assel J. and Vanrolleghem P. A. (2002). Real time control of the integrated urban wastewater system using simultaneously simulating surrogate models. *Water Science and Technology*, **45**(3), 109–116.

Melching C. S., Novotny V. and Schilling J. B. (2006). Probabilistic evaluation of ammonia toxicity in Milwaukee's Outer Harbor. *Water Science and Technology*, **53**(1), 109–116.

Morgan D. and Takyi A. (2000). Comprehensive risk assessment for developing water quality criteria for ammonia. 2000 Annual Conference – Canadian Society for Civil Engineering, June 7, 2000–June 10, 2000, London, Ontario, Canada, Canadian Society for Civil Engineering.

Novotny V. (2004). Simplified databased total maximum daily loads, or the world is log-normal. *Journal of Environmental Engineering*, **130**(6), 674–683.

Oliveira S. C. and Von Sperling M. (2008). Reliability analysis of wastewater treatment plants. *Water Research*, **42**(4–5), 1182–1194.

Palumbo J. and Maltby V. (2004). The effect of data availability and model complexity on the buxahatchee creek TMDL. *NCASI Technical Bulletin*, **878**.

Perotto E., Canziani R., Marchesi R. and Butelli P. (2008). Environmental performance, indicators and measurement uncertainty in EMS context: a case study. *Journal of Cleaner Production*, **16**(4), 517–530.

Portielje R. and Rijsdijk R. E. (2003). Stochastic modelling of nutrient loading and lake ecosystem response in relation to submerged macrophytes and benthivorous fish. *Freshwater Biology*, **48**(4), 741–755.

Qin X. S. and Huang G. H. (2009). An inexact chance-constrained quadratic programming model for stream water quality management. *Water Resources Management*, **23**(4), 661–695.

Qin X., Huang G., Chen B. and Zhang B. (2009). An interval-parameter waste-load-allocation model for river water quality management under uncertainty. *Environmental Management*, **43**, 999–1012.

Reiss R., Mackay N., Habig C. and Griffing J. (2002). An ecological risk assessment for triclosan in lotic systems following discharge from wastewater treatment plants in the United States. *Environmental Toxicology and Chemistry*, **21**(11), 2483–2492.

Ren S. (2008). Probabilistic risk assessment for zinc in some commercial dishwashing detergents: the anaerobic digestion processes in US POTWs. *Chemosphere*, **71**(3), 589–600.

Riva M., Guadagnini A., Fernandez-Garcia D., Sanchez-Vila X. and Ptak T. (2008). Relative importance of geostatistical and transport models in describing heavily tailed breakthrough curves at the Lauswiesen site. *Journal of Contaminant Hydrology*, **101**(1–4), 1–13.

Rivas A., Irizar I. and Ayesa E. (2008). Model-based optimisation of wastewater treatment plants design. *Environmental Modelling and Software*, **23**(4), 435–450.

Rossi L., Krejci V., Rauch W., Kreikenbaum S., Fankhauser R. and Gujer W. (2005). Stochastic modeling of total suspended solids (TSS) in urban areas during rain events. *Water Research*, **39**(17), 4188–4196.

Schulze F. H., Wolf H., Jansen H. W. and van der Veer P. (2005). Applications of artificial neural networks in integrated water management: fiction or future? *Water Science and Technology*, **52**(9), 21–31.

Silva N. G. M. and von Sperling M. (2008). Sensitivity analysis of non-point sources in a water quality model applied to a dammed low-flow-reach river. *Water Science and Technology*, **57**(8), 1295–1300.

Singh A. P., Ghosh S. K. and Sharma P. (2007). Water quality management of a stretch of river Yamuna: an interactive fuzzy multi-objective approach. *Water Resources Management*, **21**(2), 515–532.

Stehouwer R. C., Wolf A. M. and Doty W. T. (2000). Chemical monitoring of sewage sludge in Pennsylvania: variability and application uncertainty. *Journal of Environmental Quality*, **29**(5), 1686–1695.

Tanaka H., Asano T., Schroeder E. D. and Tchobanoglous G. (1998). Estimating the safety of wastewater reclamation and reuse using enteric virus monitoring data. *Water Environment Research*, **70**(1), 39–51.

Tata P., Lue-Hing C. and Knafl G. J. (2000). Statistical evaluation of pathogen inactivation for a conventional low-cost technology class A biosolids process. *Water Environment Research*, **72**(4), 423–431.

Torres J., Brumbelow K. and Guikema S. D. (2009). Risk classification and uncertainty propagation for virtual water distribution systems. *Reliability Engineering and System Safety*, **94**(8), 1259–1273.

Uddameri V. and Dyson B. (2007). A decision-analytic approach for designing aquaculture treatment wetlands subject to intermittent loading under uncertainty. *Water, Air, and Soil Pollution*, **186**(1–4), 297–309.

Vaes G. and Berlamont J. (1999). Emission predictions with a multi-linear reservoir model.

Van Ginneken M. and Oron G. (2000). Risk assessment of consuming agricultural products irrigated with reclaimed wastewater: an exposure model. *Water Resources Research*, **36**(9), 2691–2699.

Van Zelm R., Huijbregts M. A., Russell M. H., Jager T. and Van de meent D. (2008). Modeling the environmental fate of perfluorooctanoate and Its precursors from global fluorotelomer acrylate polymer Use. *Environmental Toxicology and Chemistry*, **27**(11), 2216–2223.

Willems P. (2006). Random number generator or sewer water quality model? *Water Science and Technology*, **54**(6–7), 387–394.

Willems P. (2008). Quantification and relative comparison of different types of uncertainties in sewer water quality modeling. *Water Research*, **42**(13), 3539–3551.

Willems P. and Berlamont J. (2001). Probabilistic modelling of the immission of water and pollution from urban drainage systems in receiving watercourses. Urban Drainage Modeling: Proceedings of the Specialty Symposium of the World Water and Environmental Resources Congress, May 20, 2001–May 24, 2001, Orlando, FL, United States, American Society of Civil Engineers.

Willems P. and Berlamont J. (2002). Probabilistic emission and immission modelling: case-study of the combined sewer – WWTP – receiving water system at Dessel (Belgium). *Water Science and Technology*, **45**(3), 117–124.

Yan J. M., Vairavamoorthy K. and Gorantiwar S. D. (2006). Contaminant transport model for unsaturated soil using fuzzy approach. *Journal of Environmental Engineering*, **132**(11), 1489–1497.

Yin T. and Stevens D. K. (2007). Regulating waste discharge with fuzzy interval effluent limitations for toxic pollutants. *Water Environment Research*, **79**(4), 414–420.

C.7 UPDATED LITERATURE 2011–2019

Barth A., Bürger R., Kröker I. and Rohde C. (2016). Computational uncertainty quantification for a clarifier-thickener model with several random perturbations: a hybrid stochastic Galerkin approach. *Computers & Chemical Engineering*, **89**, 11–26.

Batzias F. A., Salapa I. S. and Siontorou C. G. (2016). On the tradeoff between reliability and uncertainty: when combining bioreactors for wastewater treatment. Advances in environment. *Computational Chemistry and Bioscience*, **12**, 181–186.

Bedla D., Chmielowski K., Dacewicz E. and Jurik L. (2019). Effect of parametric uncertainty of selected classification models and simulations of wastewater quality indicators on predicting the sludge volume index. *Polish Journal of Environmental Studies*, **29**(2), 1–10.

Benedetti L., Belia E., Cierkens K., Flameling T., De Baets B., Nopens I. and Weijers S. (2013). The incorporation of variability and uncertainty evaluations in WWTP design by means of stochastic dynamic modeling: the case of the Eindhoven WWTP upgrade. *Water Science and Technology*, **67**, 1841–1850.

Benítez-López A., Conesa I., Douziech M., Franco A., Huijbregts M. and Zelm R. (2018). Quantifying variability in removal efficiencies of chemicals in activated sludge wastewater treatment plants – a meta-analytical approach. *Environmental Science: Processes & Impacts*, **20**, 171–182.

Borzooei S., Miranda G., Teegavarapu R., Scibilia G., Meucci L. and Zanetti M. (2019). Assessment of weather-based influent scenarios for a WWTP: application of a pattern recognition technique. *Journal of Environmental Management*, **242**, 450–456.

Chen Z. and Zayed T. (2015). An efficiency-centred hierarchical method to assess performance of wastewater treatment plants. *International Journal of Environmental Research*, **9**, 1–8.

Chen W., Lu X. and Yao C. (2015a). Optimal strategies evaluated by multi-objective optimization method for improving the performance of a novel cycle operating activated sludge process. *Chemical Engineering Journal*, **260**, 492–502.

Chen W., Lu X., Yao C., Zhu G. and Xu Z. (2015b). An efficient approach based on bi-sensitivity analysis and genetic algorithm for calibration of activated sludge models. *Chemical Engineering Journal*, **259**, 845–853.

Chhipi-Shrestha G., Hewage K. and Sadiq R. (2017). Fit-for-purpose wastewater treatment: conceptualization to development of decision support tool (I). *The Science of the Total Environment*, **607–608**, 600–612.

Cierkens K., Plano S., Benedetti L., Weijers S., Jonge J. and Nopens I. (2012). Impact of influent data frequency and model structure on the quality of WWTP model calibration and uncertainty. *Water Science Technology*, **65**(2), 233–242.

Cosenza A., Mannina G. and Viviani G. (2012). The GSA-GLUE approach for uncertainty assessment of an integrated MBR model. Proceedings of the 2012 International Congress on Environmental Modelling and Software, Sixth Biennial Meeting, Leipzig, Germany.

Cosenza A., Mannina G., Vanrolleghem P. and Neumann M. (2013). Global sensitivity analysis in wastewater applications: a comprehensive comparison of different methods. *Environmental Modelling & Software*, **49**, 40–52.

Cotman M. (2013). Sampling uncertainty of wastewater monitoring estimated in a collaborative field trial. *TrAC Trends in Analytical Chemistry*, **51**, 71–78.

Cristescu S.M., Nascu I. and Nascu I. (2015). Sensitivity analyses of an activated sludge model for a wastewater treatment plant. Proceedings: 19th International Conference on System Theory, Control and Computing (ICSTCC), 14–16 Oct., Cheile Gradistei, Romania, 595–600.

De Gisi S., Petta L., Farina R. and De Feo G. (2014). Using a new incentive mechanism to improve wastewater sector performance: the case study of Italy. *Journal of Environmental Management*, **132**, 94–106.

Djeddou M., Bachir A. and Maurice M. (2013). Determination and analysis of daily reliability level of municipal wastewater treatment plant. *Courrier du Savoir*, **17**, 39–46.

Di Maria F., Micale C. and Contini S. (2016). A novel approach for uncertainty propagation applied to two different bio-waste management options. *The International Journal of Life Cycle Assessment*, **21**, 1529–1537.

Dimitrova N. 3.3 Asymptotic Stabilization of a Bioprocess Model Involving Uncertainties. Uncertainty modeling and analysis with intervals: Foundations, tools, applications, 32–33.

Dong X., Zhang X. and Zeng S. H. (2017). Measuring and explaining eco-efficiencies of wastewater treatment plants in China: an uncertainty analysis perspective. *Water Research*, **112**, 195–207.

Dragan M. J., Zubov A. and Sin G. (2017). Methodology for plantwide design and optimization of wastewater treatment plants. Proceedings: 27th European Symposium on Computer Aided Process Engineering (ESCAPE 27), Barcelona, Spain, Oct. 1–5, 2017, **1**(40), 859–864.

Duran O. and Acevedo P. (2018). Activity based costing for wastewater treatment and reuse under uncertainty: a fuzzy approach. *Sustainability*, **10**(2260), 1–15.

Dursun M. (2016). Evaluation of wastewater treatment alternatives using fuzzy VIKOR method. *Journal of Advanced Management Science*, **4**, 333–336.

Estrada J., Kraakman B., Lebrero R. and Muñoz R. (2012). A sensitivity analysis of process design parameters, commodity prices and robustness on the economics of odour abatement technologies. *Biotechnology Advances*, **30**, 1354–1363.

Flores-Alsina X., Gernaey K. and Jeppsson U. (2012). Global sensitivity analysis of the BSM2 dynamic influent disturbance scenario generator. *Water Science and Technology*, **65**, 1912–1922.

Gimeno P., Marcé R., Bosch L., Comas J. and Corominas L. (2017). Incorporating model uncertainty into the evaluation of interventions to reduce microcontaminant loads in rivers. *Water Research*, **124**, 415–424.

Gómez T., Gémar G., Molinos-Senante M., Sala-Garrido R. and Caballero R. (2018). Measuring the eco-efficiency of wastewater treatment plants under data uncertainty. *Journal of Environmental Management*, **226**, 484–492.

Guo L., Zhao Y. and Cui F. (2015). A new fault diagnosis method based on Bayesian network model in a wastewater treatment plant of northern China. *Desalination and Water Treatment*, **57**, 1–10.

Haag F., Lienert J., Maurer M. and Reichert P. (2019). Integrating uncertainty of preferences and predictions in decision models: an application to regional wastewater planning. *Journal of Environmental Management*, **252**, 109652.

Hakanen J., Sahlstedt K. and Miettinen K. (2013). Wastewater treatment plant design and operation under multiple conflicting objective functions. *Environmental Modelling & Software*, **46**(C), 240–249.

Ilangkumaran M., Sasirekha V., Anojkumar L., Sakthivel G., Raja M., Thasapparaj R., Siddhartha C. N. S., Pallikadath N. and Kumar S. (2013). Optimization of wastewater treatment technology selection using hybrid MCDM. *Management of Environmental Quality*, **24**, 619–641.

Juznic-Zonta Z., Kocijan J., Flotats X. and Vrecko D. (2012). Multi-criteria analyses of wastewater treatment bioprocesses under an uncertainty and a multiplicity of steady states. *Water Research*, **46**, 6121–6131.

Juznic-Zonta Z., Flotats X. and Magrí A. (2014). Estimation of parameter uncertainty for an activated sludge model using Bayesian inference: a comparison with the frequentist method. *Environmental Technology*, **35**, 1618–1629.

Li B. and Stenstrom M. (2015). Practical identifiability and uncertainty analysis of the one-dimensional hindered-compression continuous settling model. *Water Research*, **90**, 235–246.

Li D., Yang H. and Liang X. (2012). Application of Bayesian networks for diagnosis analysis of modified sequencing batch reactor. *Advanced Materials Research*, **610–613**, 1139–1145.

Li D., Yuan T. and Liang X. (2017). A Methodology for modelling wastewater treatment process based on uncertainty theory. Proceedings: 6th International Conference on Energy and Environmental Protection June 29–30, Zhuhai, China.

Li Z., Qi R., An W., Mino T., Shoji T., Verstraete W., Gu J., Li S., Xu S. and Yang M. (2014). Simulation of long-term nutrient removal in a full-scale closed-loop bioreactor for sewage treatment: an example of Bayesian inference. *Frontiers of Environmental Science & Engineering*, **9**, 534–544.

Liu Y., Daoping H. and Zhifu L. (2013). A SEVA soft sensor method based on self-calibration model and uncertainty description algorithm. *Chemometrics and Intelligent Laboratory Systems*, **126**, 38–49.

Mannina G., Cosenza A. and Viviani G. (2015). Sensitivity and uncertainty analysis of an integrated membrane bioreactor model. *Desalination and Water Treatment*, **57**, 1–18.

Mannina G., Cosenza A., Gori R., Garrido-Baserba M., Sobhani R. and Rosso D. (2016). Greenhouse gas emissions from wastewater treatment plants on a plantwide scale: sensitivity and uncertainty analysis. *Journal of Environmental Engineering*, **142**, 1–11.

Martin D. and Johnson F. (2019). Incorporating uncertainty and risk into decision making to reduce nitrogen inputs to impaired waters. *Journal of Environmental Management*, **249**, 109380.

Martin C., Neumann M. B., Altimir J. and Vanrolleghem P. A. (2012). A Tool for Optimum Design of WWTPs under Uncertainty: Estimating the Probability of Compliance. Proceedings: International Congress on Environmental Modelling and Software, Sixth Biennial Meeting, Leipzig, Germany.

Mostafa K., Ghafari B., Elahe D. and Dadras P. (2013). Optimization of conventional water treatment plant using dynamic programming. *Toxicology and Industrial Health*, **31**, 1–9.

Neba F., Tornyeviadzi H., Østerhus S. and Seidu R. (2019). Self-optimizing attainable regions of the anaerobic treatment process: modeling performance targets under kinetic uncertainty. *Water Research*, **171**, 115377.

Niero M., Pizzol M., Bruun H. and Thomsen M. (2014). Comparative life cycle assessment of wastewater treatment in Denmark including sensitivity and uncertainty analysis. *Journal of Cleaner Production*, **68**, 25–35.

Obando-Montoya A. F., Zuluaga-Bedoya C. and Castro H. (2014). A comparison of non-linear state estimation techniques in activated sludge process. Proceedings: 24e congreso Argentino de Control Automatico AADECA, Buenos Aires, Argentina, pp. 24.

Pfluger A., Callahan J., Stokes-Draut J., Ramey D., Gagen S., Figueroa L. and Munakata Marr J. (2018). Lifecycle comparison of mainstream anaerobic baffled reactor and conventional activated sludge systems for domestic wastewater treatment. *Environmental Science & Technology*, **52**, 10500–10510.

Ramin E., Flores-Alsina X., Sin G., Gernaey K., Jeppsson U., Mikkelsen P. and Plosz B. (2014). Influence of selecting secondary settling tank sub-models on the calibration of WWTP models – A global sensitivity analysis using BSM2. *Chemical Engineering Journal*, **241**, 28–34.

Restrepo J., Rodriguez G. and Obregón L. (2018). Linear and nonlinear control of an upflow anaerobic sludge blanket reactor: effects of IMC-PID tuning and uncertainty. *Chemical Engineering Transactions*, **70**, 1855.

Robles Á, Ruano V., Ribes J., Seco A. and Ferrer J. (2014). Global sensitivity analysis of a filtration model for submerged anaerobic membrane bioreactors (AnMBR). *Bioresource Technology*, **158**, 365–373.

Samsudin S. (2012). Multilevel input signal for multivariable state-space identification of wastewater treatment plant. *International Journal of the Physical Sciences*, **7**, 1847–1856.

Scott C., Bailey C., Marra R., Woods G., Ormerod K. and Lansey K. (2012). Scenario planning to address critical uncertainties for robust and resilient water–wastewater infrastructures under conditions of water scarcity and rapid development. *Water*, **4**, 848–868.

Sharifi S., Murthy S., Takács I. and Massoudieh A. (2014). Probabilistic parameter estimation of activated sludge processes using Markov chain Monte Carlo. *Water Research*, **50**, 254–266.

Sin G. and Gernaey K. V. (2016). Data Handling and Parameter Estimation. Experimental Methods in Wastewater Treatment. Chapter 5.

Spiller M., Vreeburg J., Leusbrock I. and Zeeman G. (2014). Flexible design in water and wastewater engineering – definitions, literature and decision guide. *Journal of Environmental Management*, **149**, 271–281.

Talebizadeh M., Martin C., Neumann M. and Vanrolleghem P. A. (2012). Global sensitivity analysis of the ATV design guideline: which are the most important inputs for a WWTP design? In Proceedings of WEFTEC 2012, 29 September – 3 October 2012, New Orleans, USA.

Tenore A., Vieira J., Frunzo L., Luongo V. and Fabbricino M. (2018). Calibration and validation of an activated sludge model for membrane bioreactor wastewater treatment plants. *Environmental Technology*, **41**(15), 1923–1936.

Vladova R. K., Kirilova E. G. and Vaklieva-Bancheva N. G. (2018). An approach for reduction of computational complexity of a two-stage stochastic optimization problem for capturing parameters uncertainty in an ATAD system. *Chemical Engineering Transactions*, **70**, 1021–1026.

Wang X., McCarty P., Liu J., Ren N., Lee D., Yu H., Qian Y. and Qu J. (2015). Probabilistic evaluation of integrating resource recovery into wastewater treatment to improve environmental sustainability. *Proceedings: National Academy of Sciences*, **112**, 1630–1635.

Wang B., Li Z., Dai Z., Lawrence N. and Yan X. (2019). A probabilistic principal component analysis-based approach in process monitoring and fault diagnosis with application in wastewater treatment plant. *Applied Soft Computing*, **82**, 105527.

Yu M., Pan F., Lian X. and Li X. (2012). Fault diagnosis system of A2/O process based on parameter estimation and expert system. *Applied Mechanics and Materials*, **263–266**, 1647–1651.

Zeferino J., Cunha M. and Pais A. (2012). Robust optimization approach to regional wastewater system planning. *Journal of Environmental Management*, **109**, 113–122.

Zhang K., Deletic A., Page D. and McCarthy D. T. (2015). Validation framework for water-sensitive urban design treatment systems. *Water: Journal of the Australian Water Association*, **42**(6), 70–73.

Zhang J. L., Li Y. P., Huang G. H., Baetz B. W. and Lui J. (2017). Uncertainty analysis for effluent trading planning using a Bayesian estimation-based simulation-optimization modeling approach. *Water Research*, **116**, 159–181.

Zheng J., Egger C. and Lienert J. (2016). A scenario-based MCDA framework for wastewater infrastructure planning under uncertainty. *Journal of Environmental Management*, **183**, 895–908.

Appendix D

Application of uncertainty analysis methods – knowledge from other fields

D.1 INTRODUCTION

The overall objective of this chapter is to incorporate knowledge from other fields on applications of uncertainty analysis methods. The fields selected were chemical and hydrogeological (groundwater) engineering, as they share similarities with wastewater treatment (WWT). The scope and objectives of the work are as follows:

- Identify the key attributes of the chosen fields and how they compare to wastewater treatment;
- Determine how uncertainty is typically accounted for in model-based studies in the chosen fields;
- Determine the main uncertainty methods used and whether there are any novel methods not used in wastewater treatment that are transferable to the wastewater treatment field.

The research was based on a targeted literature search and review of the uncertainty analysis methods in the selected fields. Based on the experience of the team members, key areas of research were selected for review. Each team member selected an area of focus and performed an initial screening of the available research papers, followed by a detailed review of a selected set of key papers. The initial screening focused on identifying key researchers and research groups in uncertainty analysis in the selected fields. The work of these researchers formed the basis of the detailed review. A review of the different uncertainty frameworks uncovered is provided followed by the literature review.

D.2 REVIEW OF UNCERTAINTY ANALYSIS METHODS IN CHEMICAL ENGINEERING

D.2.1 Comparison of chemical engineering with wastewater treatment

D.2.1.1 Background

In general, chemical engineering deals with the production, transportation and separation of chemical products. The focus is typically on using chemical reactions to create a compound and then using

Editor(s): Evangelia Belia, Lorenzo Benedetti, Bruce Johnson, Sudhir Murthy, Marc Neumann, Peter Vanrolleghem and Stefan Weijers
doi: 10.2166/9781780401034_0175

vapour/liquid and solid/liquid separation processes to purify the product. The reactants and products are often fluids (liquids, gases, or solid slurries).

In a classic sense, chemical engineering does not involve living matter but because research in classic areas has matured (e.g., refinery operation, catalysis, batch process optimization), recent developments in society (e.g., biofuels) have pushed the boundaries of the field towards a broader set of applications. As such, chemical engineering researchers, consultants and industries are increasingly venturing into newer, unexplored areas. This includes biochemical engineering and engineering of particulate processes (e.g., powder formulations of pharmaceuticals). This also means that in these particular areas, less knowledge is available and, as a result, more uncertainty exists for such applications. However, uncertainty has not been explored to a great extent in these emerging areas.

D.2.1.2 Similarities

Wastewater treatment plants (WWTPs) share similarities with chemical plants in that they often use a series of interconnected processes to purify influent water streams using a combination of biologically mediated reactions and solid/liquid separation processes. The basic mechanistic modelling and simulation principles remain essentially the same between the two fields, with plant models being based on mass and energy balances and consisting of algebraic and/or ordinary and partial differential equations (ODE and PDE) that are solved using numerical methods. In both fields, the fluid is often assumed to be incompressible and reactors are often assumed to be well-mixed. This means that the continuity equation for fluid motion can be described by a simple volume continuity equation. When more complex analysis of fluid motion is required, both fields employ computational fluid dynamics (CFD) analysis, although this is a more recent phenomenon in WWT. CFD analysis uses numerical methods to solve the conservation equations for linear momentum, known as the Navier–Stokes equations.

Process models are used in both fields for planning, process synthesis and design, control design, comparison of alternatives, 'what-if studies', and trouble-shooting. To date, the chemical engineering field has placed more emphasis on model-based control and optimization of its plants. This has led to the use of advanced control methods such as model-predictive control (MPC) and model-based optimization of plant design and operation using techniques such as linear programming (LP), non-linear programming (NLP), mixed-integer non-linear programming (MINLP), dynamic optimization, and optimization under uncertainty.

Both industries make use of empirical models, but the chemical process industries have made greater use of data-driven models such as Box–Jenkins time-series models, neural network models, and latent variable models (i.e., principal components analysis (PCA) and partial least squares (PLS)).

D.2.1.3 Differences

The major differences between the chemical engineering field and WWT, which are of importance when considering model-based analysis, are as follows:

- Process inputs and operational variables are more certain in the chemical process industries.
- Sensors are more widely used, and the process variables are often easier to measure.
- Process inputs in the chemical process industries have less variability.
- Raw material inputs into a chemical process system are tightly controlled and buffered. Wastewater streams have a high degree of variability in flow and composition and are subject to regular daily, weekly and annual variations as well as unexpected storms and industrial discharges.
- Process inputs in the chemical process industries are often not as strongly correlated with each other.

- The raw materials are usually independently controlled in a chemical process whereas influent pollutant variables are correlated in a WWT process.
- The underlying processes operate on much faster time-scales in chemical process systems and therefore can be more easily controlled. There is more interaction between control loops in chemical process systems due to similar time-scales, which is less of a problem in WWTPs.

D.2.1.4 Comparison in the context of model-based projects

A summary of how the chemical engineering field compares with the WWT field in terms of modelling, simulation, and uncertainty issues is given in Table D.1.

It is natural for evolving scientific fields to borrow modelling, analysis, control, and optimization techniques from other fields that are further advanced. Chemical engineering and process systems engineering (PSE) have borrowed heavily from the electrical engineering, operations research, and aeronautics fields. In turn, the WWT field has drawn heavily from chemical engineering. This technology transfer has flowed from industries where the phenomena are known with greater certainty and operate on faster time-scales to industries where the phenomena are known with less certainty and operate on slower time-scales. It follows that chemical engineering is an important field to consider when looking for novel uncertainty analysis techniques, but the techniques must be tailored to the WWT field.

D.2.2 Uncertainty methods used in chemical engineering

D.2.2.1 Methods overview

Uncertainty analysis has not received as much attention in the chemical engineering literature, as compared to fields such as WWT and hydrology. This is largely because uncertainty is perceived to be a less of an issue. The perception is that process inputs and outputs are more tightly controlled and precisely known, and first principle models are generally considered to be good representations of the processes. This perception is changing, and plants are being designed for flexibility. Uncertainty in chemical engineering appears more in the context of raw material availability and pricing, changes in weather, availability of power, and product demand. For example, one source of uncertainty could be that a company may decide to change product lines in a plant depending on market demand or may decide to start producing a product that the plant was not designed for.

In chemical engineering, the largest fraction of the literature deals with finding a best model, where best is usually stated as expectation, maximum likelihood or maximum posteriori likelihood (e.g., Maria, 2004). There are also papers dealing with parameter estimation, either using a specific data set or with continually updated online data. In these studies, uncertainty is not a concern, but approximate confidence intervals or regions do follow from the Kalman filter theory or non-linear regression theory. See Vachhani *et al.* (2001), Jang *et al.* (1986), and Ramamurthi *et al.* (1993) for examples where parameters are estimated online by means of inclusion as states in the applied model or by means of recursive estimation schemes.

There are several potential reasons for the lack of attention to uncertainty analysis in chemical engineering as compared to WWT:

- Models have largely been used in the context of operation and design of potentially unsafe systems, for which constraint satisfaction (e.g., physical constraints, safety constraints) has been considered a more important problem to solve.

Table D.1 Comparison of wastewater treatment and chemical engineering fields in terms of uncertainty analysis.

Characteristic	Wastewater Treatment Field	Chemical Process Industries/Process Systems Engineering
Types of models used	Mainly continuous models Steady-state and dynamic Linear and non-linear algebraic equations Ordinary differential equations, partial differential equations, and differential/algebraic equations Some data-driven models such as latent variable models (PCA/PLS) and neural networks	Both continuous and discrete models Steady-state and dynamic Linear, non-linear, mixed-integer linear, mixed-integer non-linear algebraic equations Ordinary differential equations, PDEs, and differential/algebraic equations Data-driven models such as latent variable models (PCA/PLS) and neural networks Stochastic differential-difference equations for chemical reaction networks
Level of detail in the models	Usually lumped parameter models (spatial variation of variables and parameters ignored) so ODEs used instead of PDEs Distributed parameter models used for biofilm modelling CFD modelling becoming more prevalent Usually, model surrogate organisms not individual bacteria	Usually, lumped parameter models Distributed parameter models also used CFD modelling commonly used Stochastic simulation approach models processes at the molecular level
Sources of uncertainty	Unexpected equipment failures and operational disturbances Influent conditions (flow rate, concentrations) Variability found in influent due to diurnal, weekly, and annual patterns Variability may be well characterized but can have storm events and industrial discharges that are unexpected and unknown in advance Population growth patterns and changing water-use and climate patterns that are uncertain Future regulations	Unexpected equipment failures and operational disturbances Unexpected changes in inlet and operational conditions Changes in economic value of product Changes in availability or quality of raw materials

(Continued)

Table D.1 Comparison of wastewater treatment and chemical engineering fields in terms of uncertainty analysis (*Continued*).

Characteristic	Wastewater Treatment Field	Chemical Process Industries/Process Systems Engineering
Numerical methods used	Nonlinear algebraic solvers Differential equation solvers including stiff solvers	Linear and nonlinear algebraic solvers Differential equation solvers including stiff solvers Linear programming Nonlinear programming algorithms Mixed integer, non-linear programming methods
Data typically available	Depends on region of world Can be very limited Additional sampling campaigns typically required when doing modelling studies	Large amount of highly correlated data Data often collected with on-line instrumentation
Are models calibrated/fitted to field data? Methods used?	Yes, depends on how model used Typically use manual calibration with visual inspection	Yes, depends on how model used Typically use formal parameter estimation techniques
What are models used for?	Most common: Planning, design, control design, comparison of alternatives, and what-if studies Less common: Model-based control, process monitoring, soft sensors, plant operational advice, trouble-shooting, and optimization	Planning, scheduling, comparison of alternatives, 'what-if' studies, plant design, product design, process monitoring, soft sensors, optimization, control design and model-based control, trouble-shooting
How is uncertainty addressed within the field?	Monte Carlo analysis and error propagation analysis	Most common: Optimization under bounded uncertainty, explicit modelling of input and output disturbances, and error propagation analysis Less common: Monte Carlo analysis, polynomial chaos
What is the tolerance for uncertainty in the field?	Historically high as plants designed with large safety factors to handle the expected uncertainty in loading conditions Effluent limits are often based on averages calculated over long time periods	Low as products must meet tight quality standards and plants subject to more safety concerns due to possibility of explosions, release of toxic compounds to the atmosphere, etc.

- First principles knowledge (chemistry, thermodynamics) and extensive laboratory scale testing means that more precise identification of kinetic models for many applications can be made before full-scale modelling is attempted.
- Inputs to chemical process systems are often of a constant or well-characterized nature and process control is aimed mostly at steady-state operation in a single, well-known condition, leading to limited uncertainty and effective use of linearized models around the operating point.
- The available measurements are often easier to interpret. For example, in a refinery, temperatures, pressures, and flow rates have an explicit and unambiguous meaning. In contrast, many variables (e.g., TSS) in WWT are reflective of the process but are not easily linked with the underlying variables of interest.

Despite this, researchers in the field have developed and/or used fundamental methods to deal with uncertainty. The literature review uncovered the following methods used to account for uncertainty in a chemical engineering context:

- **Error propagation analysis**: To determine how uncertainties are propagated through an equation or a model. The main drawbacks are that its results are approximate and specific to the local solution to the parameter estimation problem.
- **Explicit modelling of input and output disturbances**: To deal with uncertainty by modelling input and output disturbances explicitly, using state estimators, like Kalman filters (Brown & Hwang, 1996; Harvey, 1989). It is often assumed that the modelling of input disturbances effectively deals with uncertainty in the parameters, so that the input disturbances are regarded as a lumped source of uncertainty, which includes parameter uncertainty.
- **Sampling methods (i.e., Monte Carlo methods) and stochastic simulation**: To eventually approximate the true distribution of uncertainties in the model inputs and parameters. One issue that is explored is whether research exists on accounting for model input correlations when using sampling techniques as part of stochastic simulation. In wastewater influents, the pollutant concentrations are correlated with each other and with other plant operational indicators (e.g., SVI) but this is often not accounted for in Monte Carlo simulation studies.
- **Bounded uncertainty**: To solve problems in production planning and scheduling, location, transportation, finance, and engineering design, using robust control algorithms and model-based decision-making. Uncertainties and disturbances are assumed to occur in a limited region. In the case of robust control, which finds control actions for the worst of possible disturbance sequences within bounds, the bounds which describe the disturbance intervals are usually used as a tuning parameter (Skogestad *et al.*, 1988; Skogestad & Postlethwaite, 1996).
- **Polynomial chaos**: To determine uncertainty propagation in predictive models. Such methods often offer a computational advantage over stochastic sampling methods (e.g., Monte Carlo), though the computational load can be still be substantial. Note that the uncertainties in the model inputs and parameters are required to be quantified ahead of time. For detailed explanations of the technique and examples of its use in chemical engineering, see Androulakis *et al.* (2006), Balakrishnan *et al.* (2002), Damian *et al.* (2002), Lovett *et al.* (2006), Mathelin *et al.* (2005), Phenix *et al.* (1998), Reagan *et al.* (2003), Reagan *et al.* (2004), Reagan *et al.* (2005), and Xiu and Karniadakis (2002).

A more thorough review of sampling methods, stochastic simulation and bounded uncertainty is provided below.

D.2.2.2 Error propagation analysis

This approach has been well documented and is widely used for calculating confidence regions for model outputs and model parameters. A typical example is in nonlinear least-squares parameter estimation, where parameter and model response uncertainty are typically assessed through the calculation of approximate joint confidence regions for the parameters and approximate confidence limits on individual model responses and parameters. The inference regions and limits are often estimated by extending linear regression theory. The model residuals are linearized using a Taylor-series expansion and analogous formulas as those used for linear regression inference regions and bands are developed.

D.2.2.3 Explicit modelling of input and output disturbances

State estimators, like Kalman filters (Brown & Hwang, 1996; Harvey, 1989), deal with uncertainty by modelling input and output disturbances explicitly. In most cases, the parameters are fixed following first principles modelling and/or model calibration. It is often assumed that the modelling of input disturbances effectively deals with uncertainty in the parameters, that is, the input disturbances are regarded as a lumped source of uncertainty which includes parameter uncertainty.

The Kalman filter is popularly interpreted as a Bayesian method for state estimation, see for example, Roweiss & Ghahramani (1999). Special challenges occur when the modelled system behaves non-linearly (nonlinear differential equations) or when the measurements are nonlinear in the state variables (nonlinear observer equations). Historically, this has been handled with the extended Kalman filter (EKF) (Becerra *et al.*, 2001; Brown & Hwang, 1996; Fotopoulos *et al.*, 1998; Harvey, 1989), which linearizes the model around the last state estimate at each time step. However, the EKF has been shown to be unstable in certain cases. In addition, the EKF cannot handle (equality and inequality) constraints very well (e.g., non-negativity of concentrations; thermodynamic balances). Particle filters (Arulampalam *et al.*, 2002; Doucet *et al.*, 2001), also called sequential Monte Carlo methods, are better suited for systems with non-linear dynamics, discrete states and non-Gaussian disturbances. Such filters are based on stochastic sampling methods, akin to Monte Carlo sampling in Bayesian integration. The downside of this approach is the computational load which comes with all stochastic sampling methods. This can be prohibitive for on-line applications. In such cases, the unscented Kalman filter (UKF) has been proposed as the better method compared to the EKF (Wan & Van der Merwe, 2000). The UKF achieves second-order accuracy (e.g., it estimates means and variances correctly) with a limited number of deterministic samples (in particle filtering the samples are numerous and chosen randomly). The UKF has been used and adapted to handle nonlinear dynamics, nonlinear observer equations, equality and inequality constraints (e.g., Mandela *et al.*, 2010; Romanenko *et al.*, 2004; Teixeira *et al.*, 2010). Note that it does not handle non-Gaussian distribution of disturbances. As a last method, the moving horizon estimation (MHE) is mentioned (Rao & Rawlings, 2002; Rao *et al.*, 2003). The MHE estimates process states at a time in the past given current and past observations. As such it is smoother, rather than a filter. Smoothers are less sensitive to disturbances because additional information is available (the observations after the considered time instant). In contrast to classic smoothers based on the Kalman filter (Kalman smoother, extended Kalman smoother), it delivers the maximum likelihood trajectory of states over a time window instead of the expected trajectory.

D.2.2.4 Sampling-based methods

Sampling methods, also known as Monte Carlo methods, use statistical sampling techniques to obtain a probabilistic approximation to the solution of a mathematical model or problem. Monte Carlo methods are discussed in the chemical engineering and PSE literature, although their use for uncertainty analysis appears to not be as prevalent as in water and wastewater engineering. Many applications of the method are not in the context of uncertainty analysis but involve the solution of mathematical problems that are difficult or impossible to solve with other numerical methods.

Gazi *et al.* (1996), Lee *et al.* (1996), Sin *et al.* (2010), Tørvi and Hertzberg (1998), Vásquez and Whiting (2000), and Vásquez *et al.* (2010) report the use of Monte Carlo methods to estimate the uncertainty in models used in the chemical process industries. Tørvi and Hertzberg studied different sampling methods including basic Monte Carlo, median Latin hypercube sampling, a Halton sequence, and a method based on Gaussian quadrature. Gaussian quadrature was found to be very accurate but not well suited to larger problems.

Gazi *et al.* (1996) use dynamic simulation for controller verification in the presence of uncertainties and employ Monte Carlo simulation to quantify the uncertainties. Qualitative reasoning techniques such as QSIM (Kuipers, 1986) are used to translate the Monte Carlo results into qualitative descriptions of the possible behaviour of the system. The qualitative descriptions are summarized in a tree structure which can be checked for interesting behaviours using computation tree logic. The approach is reported to provide the answers to qualitative questions about a system concerning its safety, reliability, and operability.

Lee *et al.* (1996) use Monte Carlo simulation in the context of dynamic chemical process simulation. Monte Carlo simulations are used to assess the uncertainty in the dynamic modelling results given the uncertainties in model inputs and parameters. The handling of discrete events such as start-up or shut-down of equipment is incorporated into their modelling approach. Statistical analysis is then used to interpret the results of the Monte Carlo analysis such scatterplots and regression analysis of the input and output samples.

Vásquez *et al.* (2010) used Monte Carlo analysis to obtain confidence limits on the output variables of a chemical process simulation models. To account for systematic errors, they used values from either the uniform distribution or another appropriate distribution when a priori information was available. Gaussian probability distributions were used to characterize random variables. The presence of systematic errors can lead to heavy tails in the probability distributions of the output variables, and Vásquez *et al.* (2010) developed a technique for estimation of the confidence intervals in these situations.

The research group led by Gintaras V. Reklaitis at Purdue has studied supply chain management and have used Monte Carlo simulation to evaluate schedule robustness under planning uncertainty (see Honkomp *et al.*, 1999; Mignon *et al.*, 1995). They have also incorporated Monte Carlo simulation into deterministic supply chain planning and scheduling models (Jung *et al.*, 2004).

The inputs to a simulation model may be correlated in some way and this can be accounted for when running Monte Carlo simulations. Wu and Tsang (2004) demonstrate the use of four different methods for generating correlated random numbers in the context of ecological modelling: Iman-Conover, standard normal transformation, normal copula, and maximum-entropy copula. The Iman and Conover (1982) method is the most well-known method and is used in software such as Crystal Ball and @Risk. The basis of the method is that independent random numbers can be transformed into correlated ones using an orthogonal transformation. The correlation matrix between the input variables is decomposed using the Cholesky decomposition. The resulting lower triangular matrix is then multiplied by the matrix of independent random numbers ($N \times k$ matrix; N sets of k independent random numbers) to produce a matrix of correlated random variables which serves as an input into the Monte Carlo simulations. In this manner, the Monte Carlo analysis is not biased by unreasonable combinations of variable values.

Sin *et al.* (2010) used Monte Carlo simulations to determine the model output uncertainty in cellulose hydrolysis models used in biofuel process design. They considered both input variable and model parameter uncertainty and considered correlation between the input parameters using the method of Iman and Conover (1982).

Vásquez and Whiting (2000) used equal probability sampling (EPS) to analyse uncertainty in thermodynamic models. They report that EPS provides more realistic results than other sampling techniques, such as Latin hypercube sampling or shifted Hammersley sampling when the model parameters are highly correlated. The EPS method involves stratifying the parameter space of the parameter estimation objective function. This is done by stratifying the probability distribution of the objective function into a number of intervals of equal probability. The inverse images of these intervals form shells of equal probability in the parameter space. A resampling scheme for each of the shells is used to ensure uniform coverage of the parameter space.

D.2.2.5 Stochastic simulation algorithm (SSA)

A variation on the Monte Carlo simulation approach, described in the physical chemistry literature, is the stochastic simulation algorithm (SSA) first discussed by Gillespie (1977). The SSA regards the time evolution of a chemical reaction system as random-walk process governed by a single differential-difference equation instead of a set of coupled ODEs. This differential-difference equation is often mathematically intractable and does not lend itself well to numerical solution (Gillespie, 1977). Instead, the stochastic simulation problem can be solved using Monte Carlo methods. The difference between the SSA and traditional Monte Carlo simulation is that in the SSA, random numbers are generated at each time step and used to determine when the next reaction will occur (how far along the next time step occurs) and what kind of reaction occurs.

The stochastic simulation approach has a firmer theoretical basis than the deterministic one as chemical reaction systems are actually discrete, stochastic processes. Molecular population levels can only change by discrete integer amounts and chemical reactions require molecular collisions which are essentially random processes when the molecules are at thermal equilibrium.

For many problems that can be represented as linear or nonlinear ODEs or PDEs, the stochastic simulation algorithm (SSA) gives results that are comparable to the numerical solution of the deterministic differential equations. The SSA produces results that have the appearance of 'noisy' solutions of the differential equations (Erban *et al.*, 2007). There are certain problems where the SSA gives results that cannot be obtained by solving the deterministic model. Examples include chemical reaction systems that have two or more stable steady states, where the SSA can predict random switching between steady states due to fluctuations in the number of molecules, or systems with self-induced stochastic resonance, where the SSA predicts oscillatory solutions.

The SSA has been applied to biological reaction–diffusion systems involving cell growth. Testing of the technique in the context of activated sludge modelling would be required to determine if the SSA has practical applications in the field of WWT. The SSA would not be used to evaluate the impact of model input or parameter uncertainty on simulation results, but instead to assess the potential for uncertain stochastic processes to impact the outcome of processes over time.

D.2.2.6 Bounded uncertainty

The concept of bounded uncertainty is used to solve chemical engineering problems where decisions are made in the presence of uncertainty, such as production planning and scheduling, transportation and location problems, finance, and engineering design. In these problems, uncertainty is considered to impact the weather, the prices of fuels, the availability of power, and the demand for resources (Sahinidis, 2004).

The study of optimization under uncertainty began with the works of Beale (1955), Bellman (1957), Charnes and Cooper (1959), Dantzig (1955), and Tintner (1955). The main approaches to optimization under uncertainty are: stochastic programming, fuzzy programming, and stochastic dynamic programming (Sahinidis, 2004).

Grossman and Sargent (1978) and Halemane and Grossman (1983) published some of the earlier studies of deterministic flexible programming approach in the field of chemical engineering. Their objective was to determine optimal process designs under uncertainty. As in stochastic programming, a two-stage strategy is used that takes advantage of the fact that control variables can be adjusted during operation to satisfy the design specifications of the plant, and that only the design of the plant remains fixed in the second stage. The strategy is intended to avoid overdesign, which can lead to non-optimal or even infeasible operation. The model used is a combination of equations and inequalities. The model variables include stage 1 design variables (plant structure and equipment sizes), stage 2 control variables that can be adjusted during operation (e.g., flow rates), state variables that describe the process, and uncertain parameters. For a given design (determined in stage 1), the next step is to solve the so-called feasibility problem to determine if the design is feasible for a realization of the uncertain parameters (stage 2). The more general problem of quantifying flexibility involves finding the maximum deviation that a given design can tolerate such that every point in the uncertain parameter space is feasible.

Dynamic programming involves the solution of multi-stage decision processes such as discrete-time systems. The problems often suffer from the curse of dimensionality as the number of state and control variables increases. See Sahinidis (2004) for a discussion of dynamic programming.

The literature review found the following areas of research and associated research papers in the field of optimization under uncertainty in chemical engineering:

(1) Optimal design of chemical plants under uncertainty:
   (a) Flexible programming with bounded uncertainties: Grossman and Sargent (1978), Halemane and Grossman (1983), Ostrovsky *et al.* (2003), Rooney and Biegler (2003), Song *et al.* (2002).
   (b) Use of a sensitivity analysis and parametric programming approach for linear process engineering problems under uncertainty: Acevedo and Pistikopoulos (1997).
   (c) Flexible programming with probability distribution functions for the uncertainties: Acevedo and Pistikopoulos (1998), Ierapetritou *et al.* (1996).
   (d) Flexible programming for dynamic systems: Mohideen *et al.* (1996)
(2) Reviews of optimization under uncertainty: Biegler and Grossmann (2004), Pistikopoulos (1995), Sahinidis (2004).
(3) Process scheduling under uncertainty: Li and Ierapetritou (2007).
(4) Supply chain design and planning under demand uncertainty: You *et al.* (2009), You and Grossmann (2010).
(5) Pharmaceutical waste management under uncertainty: Linninger and Chakraborty (2001).

D.2.2.7 Polynomial chaos expansion

Polynomial chaos is another available technique for uncertainty propagation in predictive models. The theory is based on the fact that a stochastic process can be described as an infinite combination of linear processes (spectral decomposition). Therefore, the distribution in the next time interval can be approximated based on the distributions in the current one.

The main requirement is that the uncertainties in the parameters or system inputs are quantified and are available in the form of probability density functions (PDF). Polynomial chaos techniques then proceed by expanding the available density function into an expanded basis function. Each original parameter is then converted into a set of parameters which describe its density function. Usually, a polynomial basis is used for this purpose (e.g., Hermite polynomials, Legendre polynomials). This parametric description of the density function is then propagated through a predictive model. Following this propagation, one can then recover the density function at each successive time interval. Such methods often offer a computational advantage over stochastic sampling methods (e.g., Monte Carlo), though the computational load can still be substantial. Note that the uncertainties in the model inputs and parameters are required to be quantified ahead of time. The technique is often used as part of a stochastic response surface methodology.

For detailed explanations of the technique and examples of its use in chemical engineering, see Androulakis *et al.* (2006), Balakrishnan *et al.* (2002), Damian *et al.* (2002), Lovett *et al.* (2006), Mathelin *et al.* (2005), Phenix *et al.* (1998), Reagan *et al.* (2003), Reagan *et al.* (2004), Reagan *et al.* (2005), and Xiu and Karniadakis (2002).

D.2.3 Applicability to WWT

There are some uncertainty analysis techniques used in the chemical engineering field that could be adapted for use in WWT. Optimization uncertainty appears to be a technique that could provide benefits in the area of plant design and operation. Research is needed to determine how the techniques could be best applied and whether they provide significant benefits over traditional methods. Problems based on linear programming are typically easier to solve, and may provide the most benefits with the least computational burden. The interesting aspect is that optimization under uncertainty has the potential to assist in developing plant designs that not only focus on robustness, but also flexibility and operability.

Monte Carlo methods are used in chemical engineering as in WWT, but there are some unique techniques used that warrant further study. The SSA of Gillespie (1977) may be of interest to those studying biological transformations at the cellular level. Its applicability to whole plant uncertainty analysis is likely limited. The techniques for incorporating correlation among input parameters should be tested for applicability in WWT.

Inclusion of explicit disturbances as part of state estimation or process modelling in general is a topic of interest. The concept has been applied in WWT modelling, but there may be some ideas (e.g., the disturbance models) that could be taken from chemical engineering that could improve the methodology. Running long-term dynamic simulations with realistic input disturbances is a method of uncertainty analysis and has the benefits of simplicity.

Polynomial chaos is an interesting concept and has been used by a number of researchers in chemical engineering. Further study is required to determine if whether its potential to reduce the computational load would provide benefits over existing Monte Carlo-based methods.

D.3 REVIEW OF UNCERTAINTY ANALYSIS METHODS IN HYDROGEOLOGICAL (GROUNDWATER) ENGINEERING

D.3.1 Comparison of hydrogeological engineering with WWT

D.3.1.1 Background

The modelling of hydrogeological processes shares a similar focus as WWTP modelling in that it deals with the prediction of liquid component concentrations, has dynamically varying inputs, and is concerned with the spatial/temporal variability of parameters.

Hydrogeological (groundwater) modelling focuses on the following:

- Modelling of the flow of liquids (of various densities) in a partially or fully saturated porous heterogeneous media;

- Transport of contaminants in porous or fractured heterogeneous media via advective, diffusive or dispersive processes; and
- Transformative processes during transport (e.g., biodegradation, sorption to media surfaces, volatilization, dissolution of dense phase liquids).

D.3.1.2 Comparison in the context of model-based projects

Despite some of the similarities between hydrogeological processes and WWTPs, the associated uncertainty is often fundamentally different than WWTP uncertainty. The major sources of uncertainty in hydrogeological modelling are in the characteristics of the systems (i.e., aquifers) being modelled. Aquifers can be difficult to model for the following reasons:

- They are buried underground out of plain sight, and often not easily observable/measurable,
- Are spatially heterogeneous,
- Have boundaries (in three dimensions) that are often unknown,
- Can hydraulically connect to other aquifers,
- Can be difficult and expensive to sample, and
- Conceptual structural models are often assumed to be 'correct' if they can be calibrated, and subsequently can be difficult to disprove once accepted.

A summary of the differences between groundwater and WWTP models is provided in Table D.2.

Table D.2 Comparison of groundwater and WWTP models.

Groundwater Models	WWTP Models
System size, boundaries, composition often unknown/unmeasurable	Well-defined, measurable tanks, with known hydraulic connectivity
Significant spatial heterogeneity (e.g., porosity, conductivity)	Effective tank volumes often well known.
Potentially significant discontinuities from depositional origins (e.g., discrete boundaries between clay and silt layers) not easily identified	Biofilm structure not as well known. Aerated/non-aerated zones are often defined by design, and often well-known and/or measurable
Underlying model structure definition (e.g., aquifer thickness) unlikely to change temporally, even if not well known.	Some system definition can change temporally (e.g., tank volumes in SBRs) but more likely to be known/measurable.

As in our review of the chemical engineering field, the focus of this report is on methods for accounting for uncertainty in model-based projects. Common uncertainty analysis methods used in this field include:

- Calculation of approximate confidence regions and limits around parameters and model outputs as part of parameter estimation.
- Stochastic simulation in classic and Bayesian frameworks.

Due to the nature of the systems modelled, application of the above techniques can be computationally intensive. In addition, aquifer systems can be very non-linear and discontinuous, limiting the use of approximate confidence regions and intervals.

D.3.2 Uncertainty methods used in hydrogeological engineering

D.3.2.1 Model structure uncertainty

It is a widespread challenge in science to develop and use models without any explicit fundamental philosophy. The lack of interdisciplinary knowledge, particularly in mathematics, is often the reason for a scientist, working for instance in the field of hydrology, to develop solutions that are numerically incorrect.

Other limitations can be the constraints of current knowledge, computing capabilities and observational technologies. In the Bayesian approach to inverse problems, prior estimates of model parameter distributions are adjusted on the basis of a likelihood measure that can demonstrate how well a model predicts the available set of observations to calculate a posterior distribution of parameters. In this approach, likelihood functions can be used to approximate parameter values. With increased availability of experimental data, the use of these likelihood functions can potentially lead to reduced uncertainty in parameter values.

This probabilistic framework is considered as one of the only potential approaches to address model uncertainty. However, one cannot treat the entire range of sources of uncertainty in an aggregate form because, for instance, model structure uncertainties are nonlinear, non-stationary and non-additive (Beven, 2006). Thus, they cannot be accounted for by the likelihood-based approach.

Bayes methods have a number of advantages, one of them being that different model structures can be compared and combined. For many hydrological models, and presumably for WWTP models, the definition of a formal likelihood measure can lead to misleading results if the assumptions on which it is based upon are not valid. That is, it is questionable whether the various sources of uncertainty can be represented adequately by a formal uncertainty structure, which defines the appropriate likelihood function. For most of the hydrological models, being subject to uncertainty derived from input, boundary conditions and model structure uncertainty, it is generally only possible to approximately represent the complexity of the uncertainties. As a result, the likelihood function will only be an approximate, and the resulting parameter estimates may well be biased.

From a statistical viewpoint, one can argue that model structure uncertainty can be represented by means of a model discrepancy function. O'Hagan and Oakley (2004) suggest that the complexity of observed uncertainty series in most of the hydrogeological problems does not directly imply that one should not use a formal likelihood approach, but that finding an appropriate likelihood function may require some more detailed assessment of various tools. Then, it is noteworthy that the problem is that the more complex the model of the uncertainty used, the more the number of statistical parameters that must be estimated. In some hydrogeological cases with complex uncertainty structures, the method can still be applied using a transformation of the modelling uncertainty so that the assumptions of a simple formal likelihood measure are more closely approximated. Other less formal methods are also used in the area, in cases where the choice of a formal likelihood would be incoherent.

D.3.2.2 Model identifiability (equifinality)

When using statistical methods for model calibration (Pareto optimal set approach), there is an underlying presumption that the experimental data are adequate to identify an optimal model structure (or Pareto optimal set of models). This should be the case for Bayesian methods that aim to identify the complete multi-parameter posterior distribution. Beven and Young (2003) argue that oversimplification of likelihood functions often leads to this result in hydrogeological simulations. In general, there is no such thing as the correct model structure, and one can only find true model parameters for a given model structure.

An acceptance of model equifinality is, in part, the recognition of possible model structure and input data uncertainties. It means that the formal model of equations that are used to represent the hydrogeological systems may be, at times, a poor approximation of the relevant processes. It also accounts for the possibility that, even if there was a correct model structure, it may be difficult to specify accurately all the boundary conditions (e.g., shape and size of an aquifer) required to run the correct model. There is, however, not much one can do about model structure uncertainty in most hydrogeological modelling cases since, if there were obvious improvements to be made, then there would be no reason why this should not be done (limit: computational feasibility).

D.3.2.3 Conceptual model uncertainty

There are many studies in the hydrogeological literature that report on the difficulties of finding a single 'true' model to represent a process. It seems that model structure uncertainty is something that is endemic to most of the models in the field. Beven and Freer (2001), and Rojas *et al.* (2010), both illustrate the tendency for practitioners to assume that well calibrated models can be accepted as 'true' interpretations of a system structure.

Refsgaard (2006) demonstrate an example of how model structure can introduce considerable variance in model results. Results from a hydrogeologic modelling exercise in the County of Copenhagen, Denmark, are used to illustrate the effects of different model structures on final model results. Five different consulting companies were asked to develop models of groundwater contamination risk in a 175 km^2 area west of Copenhagen. Each of the consultants was well-respected in the industry, with considerable experience in modelling contaminant flow. Each of the consultants took a different approach to the model structure, with some using a criteria-based method for risk assessment, while others used hydrological models of varying levels of complexity. In each case, the consultant assumed the underlying model structure was suitably correct, based upon past experience. The results from the five consultants differed substantially, even though all five were using the same raw data (and therefore the same data uncertainty) indicating the major source of uncertainty in model predictions was due to differences in model structure.

D.3.2.4 Model conditioning

Alternative approaches to model calibration are required to account for the effects of model structure and data uncertainty – again, despite the fact that some of these uncertainties cannot be represented explicitly. One alternative option in the field of hydrology is to identify a set of equations that derive acceptable uncertainty in the range of the available data – a process called as model conditioning. In hydrogeology, such approaches have generally been based on some form of Monte Carlo sampling from the population of feasible models. Based on the simulation results obtained with each model in the selected population, a qualitative or quantitative assessment is undertaken as to whether a particular model is accepted/rejected as behavioural. This is the basis for the generalized likelihood uncertainty estimation (GLUE) methodology, which was used, for the first time, by Beven and Binley (1992) in an application to a hydrological model.

The GLUE methodology, used with a formal uncertainty model and likelihood, infers essentially identical results to that obtained using a formal Bayesian likelihood approach. It has been noted that, for forward simulations, a set of behavioural models can be used to provide a prediction range of model variables as conditioned on the process of model evaluation. The fuzzy or probabilistic weights associated with each model can be used to weigh the model simulation to reflect how well that particular model has performed in the past.

Traditional use of the GLUE methodology in groundwater modelling supplements the forward propagation of parametric uncertainty (and/or spatial variability of parameters) through the model with posterior information on the level of correspondence between model predictions and field observations. The posterior analysis assists in the development of uncertainty bounds for each input parameter. Because these measurements of uncertainty use measured field observations in the analysis of input parameter uncertainty, the analysis is restricted to only those systems for which suitable data can be observed. This diminishes the usefulness of traditional GLUE analyses to only model evaluation, and not situations where predictive modelling is carried out (Hassan *et al.*, 2008).

To address this issue, several alternative variations on the GLUE methodology have been proposed. Rojas *et al.* (2008) combine GLUE methodology with Bayesian model averaging (BMA) to account for the uncertainty associated with the choice of model structure. In addition to propagating parametric uncertainty, the variability associated with models of different structures is incorporated into the analysis by modelling the system with a group of plausible models. A hypothetical example of the prediction of groundwater flow and head distribution within an aquifer is used to illustrate that while some predictions varied considerably among the three models used, a comparison of predictions to the observed data was unable to distinguish between models. When considering a combined prediction using BMA, the combined prediction was more conservative than individual predictions from each model. Most importantly, 30% of the total uncertainty was associated with the choice of model structure.

Rojas *et al.* (2010) applied the above multi-model approach to a real aquifer system in the Walenbos Nature Reserve area in Belgium. Using a combination of GLUE and BMA, the authors modelled the flow through the aquifer with three different models (of different levels of geological knowledge), with associated input parameter distributions. Some parameters were common amongst the three models, while others were unique. The concept of equifinality, as defined above, states that the combination of many alternative models and parameter sets can produce equally good results when compared to limited observations (Beven & Freer, 2001). In this study, GLUE analysis provided weights for each conceptual model, and the results were combined via BMA. This approach, which no longer relies on a single parameter set or conceptual model, was applied to the hydrogeologically complex aquifer system to model the hydraulic budget under various recharge scenarios.

A key conclusion of the study illustrates that typically limited observational data (in this case, observations of head in various locations in the aquifer) often cannot discriminate between conceptual models, as shown by small differences in posterior model probabilities. An additional important conclusion is that despite these small differences, the *predictive* distributions were different in shape and spread among the alternative conceptual models and scenarios analysed. The authors emphasize the point that relying on a single conceptual model driven by a particular simulation scenario will likely produce 'biased and under-dispersive estimations of predictive uncertainty'.

D.3.3 Applicability to wastewater treatment

For parameter uncertainty analysis, simple Bayesian approaches used in the hydrogeological sciences are applicable to WWT modelling, even though wastewater systems are generally more well defined than groundwater aquifers. The lack of easily observable information on the completeness of mixing in biological reactors is an example of a source of uncertainty that is analogous to the lack of information on aquifer heterogeneity in the field. Approaches to assessing this unknown and quantifying the predictive uncertainty in the hydrogeological field would be transferrable to WWT modelling.

In particular, the GLUE and BMA methods would be useful to assist in quantifying the uncertainty associated with model structure. While model structure uncertainty is generally limited to describing the physical characteristics of the aquifer systems in groundwater modelling (which is analogous to the mixing issue in wastewater modelling), there is an opportunity in the wastewater field to extend the multi-model approach to the biological model structure as well.

In groundwater modelling, the active physical processes (advection, sorption/desorption, volatilization, etc.) are well understood and well described by mathematical relationships. The major source of uncertainty comes from the spatial heterogeneity of the physical properties of the aquifer. By contrast, the active biological processes in an activated sludge tank are more complex and less well described mathematically (e.g., hydrolysis, biomass growth, etc.) than the physical characteristics of the system (e.g., hydraulic connectivity, effective tank volume, etc.). The opportunity exists to apply the multi-model approach to account for conceptual model uncertainty in the biological model specifically. With a set of activated sludge models already established in the industry (ASM1, ASM2d, ASM3, commercial simulator default models), a WWT system could be easily modelled with several biological models to provide an evaluation of the uncertainty associated with the conceptual model structure.

REFERENCES

References for uncertainty analysis in chemical engineering

Acevedo J. and Pistikopoulos E. N. (1997). A multiparametric programming approach for linear process engineering problems under uncertainty. *Industrial & Engineering Chemistry Research*, **36**, 717–728.

Acevedo J. and Pistikopoulos E. N. (1998). Stochastic optimization-based algorithms for process synthesis under uncertainty. *Computers & Chemical Engineering*, **22**(4/5), 647–671.

Androulakis I. P., Grenda M. G., Barckholtz T. A. and Bozzelli J. W. (2006). Propagation of uncertainty in chemically activated systems. *AICHE Journal*, **52**(9), 3246–3256.

Arulampalam S., Maskell S., Gordon N. and Clapp T. (2002). A tutorial on particle filters for online nonlinear/non-Gaussian Bayesian tracking. *IEEE Transactions Signal Processing*, **50**, 174–188.

Balakrishnan S., Georgopoulos P., Banerjee I. and Ierapetritou M. (2002). Uncertainty considerations for describing complex reaction systems. *AIChE Journal*, **48**(12), 2875–2889.

Beale E. M. L. (1955). On minimizing a convex function subject to linear inequalities. *Journal of the Royal Statistical Society*, **17B**, 173–184.

Becerra V. M., Roberts P. D. and Griffiths G. W. (2001). Applying the extended Kalman filter to systems described by nonlinear differential-algebraic equations. *Control Engineering Practice*, **9**, 267–281.

Bellman R. E. (1957). Dynamic Programming. Princeton University Press, Princeton, PA.

Biegler L. T. and Grossmann I. E. (2004). Retrospective on optimization. *Computers & Chemical Engineering*, **28**, 1169–1192.

Brown R. G. and Hwang P. Y. C. (1996). Introduction to Random Signals and Applied Kalman Filtering with Matlab Exercises and Solutions, 4th edn, published by Wiley, USA.

Charnes A. and Cooper W. W. (1959). Chance-constrained programming. *Management Science*, **6**, 73–79.

Damian V., Sandu A., Damian M., Potra F. and Carmichael G. R. (2002). The kinetic PreProcessor KPP – a software environment for solving chemical kinetics. *Computers and Chemical Engineering*, **26**, 1567–1579.

Dantzig G. B. (1955). Linear programming under uncertainty. *Management Science*, **1**, 197–206.

Doucet A., De Freitas N. and Gordon N. J. (2001). Sequential Monte Carlo Methods in Practice. Springer, New York, NY. https://doi.org/10.1007/978-1-4757-3437-9 Springer.

Erban R., Chapman J. and Maini P. (2007). A practical guide to stochastic simulations of reaction-diffusion processes. *arXiv*, **0704**, 1908 [q-bio.SC].

Fotopoulos J., Georgakis C. and Stenger Jr., H. G. (1998). Use of tendency models and their uncertainty in the design of state estimators for batch reactors. *Chemical Engineering and Processing*, **37**, 545–558.

Gazi E., Ungar L. H., Seider W. D. and Kuipers B. J. (1996). Automatic analysis of Monte-Carlo simulations of dynamic chemical plants. *Computers & Chemical Engineering*, **20**(Suppl.), S987–S992.

Gillespie D. T. (1977). Exact stochastic simulation of coupled chemical reactions. *Journal of Physical Chemistry*, **81**(25), 2340–2360.

Grossman I. E. and Sargent R. W. H. (1978). Optimum design of chemical plants with uncertain parameters. *AIChE Journal*, **24**(6), 1021–1028.

Halemane K. P. and Grossmann I. E. (1983). Optimal process design under uncertainty. *AIChE Journal*, **29**(3), 425–433.

Harvey A. C. (1989). Forecasting, Structural Time Series Models and the Kalman Filter. Cambridge University Press, UK.

Honkomp S. J., Mockus L. and Reklaitis G. V. (1999). A framework for schedule evaluation with processing uncertainty. *Computers & Chemical Engineering*, **23**, 595–609.

Ierapetritou M. G., Acevedo J. and Pistikopoulos E. N. (1996). An optimization approach for process engineering problems under uncertainty. *Computers & Chemical Engineering*, **20**(6/7), 703–709.

Iman R. L. and Conover W. J. (1982). A distribution-free approach to inducing rank order correlation among input variables. *Communications in Statistics – Simulation and Computation*, **11**, 311–334.

Jang S., Joseph B. and Mukal H. (1986). Comparison of two approaches to on-line parameter and state estimation of nonlinear systems. *Industrial & Engineering Chemistry Process Design and Development*, **25**, 809–814.

Jung J. Y., Blau G., Pekny J. F., Reklaitis G. V. and Eversdyk D. (2004). A simulation based optimization approach to supply chain management under design uncertainty. *Computers & Chemical Engineering*, **28**, 2087–2106.

Kuipers B. J. (1986). Qualitative simulation. *Artificial Intelligence*, **29**, 289–338.

Lee K. W., Kang J. L., Choi S. H. and Yoon E. S. (1996). Stochastic dynamic simulation of chemical processes with changing uncertainties. *Computers & Chemical Engineering*, **20**, Suppl., S557–S562.

Li Z. and Ierapetritou M. G. (2007). Process scheduling under uncertainty using multiparametric programming. *AIChE Journal*, **53**(12), 3183–3203.

Linninger A. A. and Chakraborty A. (2001). Pharmaceutical waste management under uncertainty. *Computers and Chemical Engineering*, **25**, 675–681.

Lovett T. E., Ponci F. and Monti A. (2006). A polynomial chaos approach to measurement uncertainty. *IEEE Transactions on Instrumentation and Measurement*, **55**(3), 729–736.

Mandela R. K., Rengaswamy R., Narasimhan S. and Sridhar L. N. (2010). Recursive state estimation techniques for nonlinear differential algebraic systems. *Chemical Engineering Science*, **65**(16), pp. 4548–4556.

Maria G. (2004) A review of algorithms and trends in kinetic model identification for chemical and biochemical systems, *Chemical and Biochemical Engineering Quarterly*, **18**(3), 195–222.

Mathelin L., Hussaini M. Y. and Zang T. A. (2005). Stochastic approaches to uncertainty quantification in CFD simulations. *Numerical Algorithms*, **38**(1), 209–236.

Mignon D. J., Honkomp S. J. and Reklaitis G. V. (1995). A framework for investigating schedule robustness under uncertainty. *Computers & Chemical Engineering*, **19**, Suppl., S557–S562.

Mohideen M. J., Perkins J. D. and Pistikopoulos E. N. (1996). Optimal design of dynamic systems under uncertainty. *AIChE Journal*, **42**(8), 2251–2271.

Ostrovsky G. M., Datskov I. V., Achenie L. E. K. and Volin Y. M. (2003). Process uncertainty: case of insufficient process data at the operation stage. *AIChE Journal*, **49**(5), 1216–1232.

Phenix B. D., Dinaro J. L., Tatang M. A., Tester J. W., Howard J. B. and McRae G. J. (1998) Incorporation of parametric uncertainty into complex kinetic mechanisms: application to hydrogen oxidation in supercritical water. *Combustion and Flame*, **112**, 132–146.

Pistikopoulos E. N. (1995). Uncertainty in process design and operations. *Computers & Chemical Engineering*, **19**, S553–S563.

Ramamurthi Y., Sistu P. and Bequette B. (1993). Control relevant dynamic data reconciliation and parameter estimation. *Computers & Chemical Engineering*, **17**, 41–59.

Reagan M. T., Najm H. N., Ghanem R. G. and Knio O. M. (2003). Uncertainty quantification in reacting flow simulations through non-intrusive spectral projection. *Combustion and Flame*, **132**, 45–55.

Reagan M. T., Najm H. N., Debusschere B. J., Le Maître O. P., Knio O. M. and Ghanem R. G. (2004). Spectral stochastic uncertainty quantification in chemical systems. *Combustion Theory and Modeling*, **8**, 607–632.

Reagan M. T., Najm H. N., Pébay P. P., Knio O. M. and Ghanem R. G. (2005). Quantifying uncertainty in chemical systems modeling. *International Journal of Chemical Kinetics*, **37**(6), 368–382.

Romanenko A., Santos L. O. and Afonso P. A. F. N. A. (2004). Unscented Kalman filtering of a simulated pH system. *Industrial & Engineering Chemistry Research*, **43**(23), 7531–7538.

Rooney W. C. and Biegler L. T. (2003). Optimal process design with model parameter uncertainty and process variability. *AIChE Journal*, **49**(2), 438–448.

Roweiss S and Ghahramani Z. (1999). A unifying review of linear Gaussian models. *Neural Computation*, **11**, 305–345.

Sahinidis N. V. (2004). Optimization under uncertainty: state-of-the-art and opportunities. *Computers and Chemical Engineering*, **28**, 971–983.

Sin G., Meyer A and Gernaey K. V. (2010). Assessing reliability of cellulose hydrolysis models to support biofuel process design – identifiability and uncertainty analysis. *Computers and Chemical Engineering*, **34**, 1385–1392.

Skogestad S. and Postlethwaite I. (1996). Multivariate Feedback Control. John Wiley & Sons, NJ, USA.

Skogestad S., Morari M. and Doyle J. C. (1988) Robust control of ill-conditioned plants: high-purity distillation. *IEEE Transactions on Automatic Control*, **33**, 1092–1105.

Song J., Stephanopoulos G. and Green W. H. (2002). Valid parameter range analyses for chemical reaction kinetic models. *Chemical Engineering Science*, **57**, 4475–4491.

Teixeira B. O. S., Torres L. A. B., Aguirre L. A. and Bernstein D. S. (2010). On unscented Kalman filtering with state interval constraints. *Journal of Process Control*, **20**, pp. 45–57.

Tintner G. (1955). Stochastic linear programming with applications to agricultural economics. In: Proceedings of the Second Symposium in Linear Programming, H. A. Antosiewicz (ed.), National Bureau of Standards, Washington, DC, pp. 197–228.

Vachhani P., Rengaswamy R. and Venkatasubramanian V. (2001). A framework for integrating diagnostic knowledge with nonlinear optimization for data reconciliation and parameter estimation in dynamic systems. *Chemical Engineering. Science*, **56**, 2133–2148.

Vásquez V. R. and Whiting W. B. (2000). Uncertainty and sensitivity analysis of thermodynamic models using equal probability sampling (EPS). *Computers & Chemical Engineering*, **23**, 1825–1838.

Vásquez V. R., Whiting W. B. and Meerschaert M. M. (2010). Confidence interval estimation under the presence of non-Gaussian random errors: applications to uncertainty analysis of chemical processes and simulation. *Computers & Chemical Engineering*, **34**, 298–305.

Wan E. A. and Van der Merwe R. (2000). The unscented Kalman filter for nonlinear estimation. *Proceedings Symposium. Adaptive Systems Signal Processing Communications and Control*. Lake Louise, AB, Canada, October 2000.

Wu F. and Tsang Y. (2004). Second-order Monte Carlo uncertainty/variability analysis using correlated model parameters: application to salmonid embryo survival risk assessment. *Ecological Modelling*, **177**, 393–414.

Xiu D. and Karniadakis G. E. (2002). The Wiener–Askey polynomial chaos for stochastic differential equations. *SIAM Journal on Scientific Computing*, **24**(2), 619–644.

You F. and Grossmann I. E. (2010). Integrated multi-echelon supply chain design with inventories under uncertainty: MINLP models, computational strategies. *AIChE Journal*, **56**(2), 419–439.

You F., Wassick J. M. and Grossmann I. E. (2009). Risk management for a global supply chain planning under uncertainty: models and algorithms. *AIChE Journal*, **55**(4), 931–945.

References for uncertainty analysis in hydrogeological engineering

Beven K. J. (2006). A manifesto for the equifinality thesis. *Journal of Hydrology*, **320**, 18–36.

Beven K. and Binley A. (1992). The future of distributed models – model calibration and uncertainty prediction. *Hydrological Processes*, **6**(3), 279–298.

Beven K. and Freer J. (2001). Equifinality, data assimilation, and uncertainty estimation in mechanistic modelling of complex environmental systems using the GLUE methodology. *Journal of Hydrology*, **249**, 11–29.

Beven K. and Young P. (2003). Comment on 'Bayesian recursive parameter estimation for hydrologic models' by M. Thiemann, M. Trosset, H. Gupta and S. Sorooshian, *Water Resources Research*, **39**(5), Article Number: 1116, Com 1-4.

Hassan A., Bekhit H. and Chapman J. (2008). Uncertainty assessment of a stochastic groundwater flow model using GLUE analysis. *Journal of Hydrology*, **362**, 89–109.

O'Hagan A. and Oakley J. E. (2004). Probability is perfect, but we can't elicit it perfectly. *Reliability Engineering and System Safety*, **85**(1–3), 239–248.

Rao C. V. and Rawlings J. B. (2002). Constrained process monitoring: Moving-Horizon approach. *AIChE Journal*, **48** (1), 97–109. DOI: 10.1002/aic.690480111.

Rao C. V., Rawlings J. B. and Mayne D. Q. (2003). Constrained State Estimation for Nonlinear Discrete-Time Systems: Stability and Moving Horizon Approximations. *IEEE Trans. on Auto. Cont*, **48**(2), 246–257.

Refsgaard J. (2006). A framework for dealing with uncertainty due to model structure error. *Advances in Water Resources*, **29**, 1586–1597.

Rojas R., Feyen L. and Dassargues A. (2008). Conceptual model uncertainty in groundwater modelling: Combining generalised likelihood uncertainty estimation and Bayesian model averaging. *Water Resources Research*, **44** (12), 16, Article W12418.

Rojas R., Samalie K., Peeters L., Batelaan O., Feyen L. and Dassargues A. (2010). Application of a multimodel approach to account for conceptual model and scenario uncertainty in groundwater modelling. *Journal of Hydrology*, **394**, 416–435.

Tørvi H. and Hertzberg T. (1998). Methods for evaluating uncertainties in dynamic simulation – A comparison of performance. *Computers & Chemical Engineering*, **22**, Supplement 1, 15, S985–S988.

Appendix E

Current practices in different countries

E.2 CURRENT PRACTICE IN NORTH AMERICA

E.2.1 Planning phase

In North America, planning for a wastewater treatment infrastructure project is almost universally done at the owner level. The development of the plan itself is either done by the owner or by a designer/engineer hired by the owner to work with them on developing an appropriate plan. In either situation, the risk for the planning decisions rests with the owner.

At the planning level, cost estimation used for budgeting purposes for capital improvements should follow American Society of Testing and Materials (ASTM) standard E2516 (ASTM, 2011). This standard developed a five-tiered cost estimation matrix based on the degree of project definition. As the degree of project definition increases, the accuracy of the estimate will be more refined. At the planning-level stage, this matrix defines the level of accuracy to be between – 60% and +120% of the cost estimate developed.

The following sections describe how different project delivery methods transfer risk.

E.2.2 Design–bid–build contracts

E.2.2.1 Preliminary design

The design–bid–build approach is very common in North America. Historically, in North America, design criteria for processes have been selected from one of a number of sources including local regulatory requirements or industry-accepted design standards that are generally published for reference when undertaking the process design and operation of the facility. Some examples of these design standards include:

- Water Environment Federation Manual of Practice 8 (WEF MOP-8, 2009)
- Wastewater Engineering: Treatment and Resource Recovery 5th Edition (Metcalf & Eddy Inc. *et al.*, 2013); (Tchobanoglous *et al.*, 2003)

Editor(s): Evangelia Belia, Lorenzo Benedetti, Bruce Johnson, Sudhir Murthy, Marc Neumann, Peter Vanrolleghem and Stefan Weijers
doi: 10.2166/9781780401034_0195

- EPA Nitrogen Control Manual (USEPA, 1993).
- EPA Phosphorus Removal Design Manual (USEPA, 1987)
- Biological Wastewater Treatment (Grady *et al.*, 2011)
- Methods for Wastewater Characterization in Activated Sludge Modeling (Melcer *et al.*, 2003)
- WERF/CRTC Protocols for Evaluating Secondary Clarifier Performance (Wahlberg, 2004)
- Virginia's Sewage Collection and Treatment Regulations (Virginia DEQ, 2008)
- Biological Nutrient Removal (BNR) Operation in Wastewater Treatment Plants (WEF MOP-29, 2005)
- Great Lakes Upper Mississippi River Board, Recommended Standards for Wastewater Treatment Facilities (Ten State Standards) (GLUMRB, 2014).

North American engineers place safety in their design in a few key process variables, most of which describe the most important sources of uncertainty, for example: influent flows and mass loads, SRT, SVI, overflow rates, denitrification rates and the design of the process air system.

E.2.2.2 Detailed design and construction

During detailed design for conventional design–bid–build projects, risks are mostly assumed by the designer/engineer who has to follow through the concepts of the preliminary design while preparing contract documents that can be built from. The owner does accept some risk accepting the design criteria presented in the preliminary design report, type of equipment that is provided, and operational procedures.

The development of tight contract documents and specifications is essential for the successful management of risk during construction. Reliability and redundancy standards are also used to reduce the risk of failure due to individual unit processes being out of service either due to mechanical failure or maintenance. The EPA and some states have developed standards based on the sensitivity of the receiving stream and the criticality of each unit process or piece of equipment. Typical requirements include:

- A minimum of two aeration basins of equal volume;
- Multiple pumps that can pump peak flows with the largest pump out of service; or
- Multiple units with the capacity to treat a certain percentage of design flows.

In the design–bid–build approach, no flexibility is given to the contractor to change the amount of redundant equipment, so the risk of adequate reliability falls upon the owner when the plant goes into operation.

E.2.2.3 Operation

Design–bid–build places all the operational risks upon the owner. There are some initial risks to the designer/engineer that the process design will perform as predicted, but once the plant resumes normal operation after the project is completed, the designer/engineer and contractor are no longer involved.

E.2.3 Design–build contracts

E.2.3.1 Preliminary design

In the design–build contract types, there may be an emphasis at the preliminary design stage to introduce additional risk in the design, which will be borne onto the owner during facility operations, in order to reduce construction costs. The following sources of uncertainty during preliminary design were identified earlier in this chapter:

- Variability in influent flows and loads;
- Selection of effluent design criteria;
- Selection of aerobic solids retention time;
- Selection of design sludge volume index (SVI);
- Denitrification rates;
- Considerations in the design of the process air system.

Examples of potential items which would introduce more risk to the owner would include a reduction in aerobic SRT (thereby reducing the aerobic bioreactor volume), use of a lower SVI (thereby reducing clarifier sizing), and a reduction in process air requirements (which would decrease blower or mechanical surface aerator size, compressed air pipe size (if applicable), and number of diffusers (if applicable)).

A second source of risk that is addressed in this stage that could be shifted to the owner during design would be the cost to operate the facility. Examples of potential items that could potentially provide capital cost savings during construction but increase operational costs include the use of chemical phosphorus removal in lieu of biological phosphorus removal and use of aeration systems that are not as efficient (mechanical aerators or course bubble diffused air).

Development of a strong design criteria package

To prevent undesirable items in the delivered plant, it is critical in a design–build project to have a detailed contract–design basis package that is used to specify the owner's goals for the project. However, the design package should consider a design change path for the contractor to allow them to propose cost savings ideas in a structured and thought-out manner.

At this stage of the design process, the owner will (most likely) hold a contract with a third-party engineer who is responsible for preparing a design criteria package, which may include a preliminary design of the facility along with minimum standards the facility must be designed around (the 'owner's engineer'). This will include design criteria for process mechanical, structural, electrical, instrumentation, HVAC (heating ventilation and air conditioning) and plumbing, and architectural codes and standards.

The development of a strong design criteria package by the third-party engineer can specify minimum requirements for items above such as minimum aerobic SRT, minimum clarifier solids/hydraulic loading rates, or standard oxygen requirements to prevent the design from becoming overly aggressive and making the owner more risk adverse during operations to save money during design and construction. In order to realise the benefits of this approach however, that is, reduced costs and risks to the owner, careful consideration is needed in the development of the design package to balance the true minimum requirements and the costs of being overly conservative.

E.2.3.2 Detailed design and construction

In the case of a design–build contract, the owner mitigates project risk by having both the designer/engineer and the contractor under one contract. This eliminates the so-called 'finger pointing' during the construction process. It also, however, provides the owner less control over the aspects of the project unless these items are specified in the design criteria package, which typically accompanies a design–build proposal. In this case, the owner is taking on risk by not having as much control over day-to-day decisions (such as ensuring that a specific manufacturer of a unit is provided) as long as the contract requirements are met and the price to perform the work remains the same. For example, the design–build may use alternative manufacturers or materials of construction as long as the minimum specified requirements in the design criteria package are met.

In design–build, the risks during detailed design and construction lie entirely with the design–build team to determine the level of design drawings needed to proceed to construction. There is some flexibility in this approach since it is possible to change the design during construction, if needed, to address issues that were not originally considered. In the traditional design–bid–build approach such changes normally result in additional change orders from the contractor that add to project costs.

The design–build contractor is fully responsible for the costs of the project, so the contract must be clear about where the project risks are and the limits on those risks. The goal of the approach would be to best allocate risks to where they are best handled, either at the owner level, or at the design–build level. Since it is not possible to have a 'perfect' contract, the design/builder will normally require supervision by the owner/owner's engineer to ensure that the intent of the contract is met.

Communication

Design–build eliminates some of the design–bid–build risks related to communication between the engineer and the contractor but makes the communication between the owner/their consultant and the design/builder critical. The owner and their consultant have less control over the product at this stage than is normal in a design–bid–build delivery, so communications need to be held in light of the contract language, which reduces the ability of the owner/owner's engineer to influence the design.

E.2.3.3 Operation

As in design–bid–build, the risks during operation are almost entirely upon the owner upon commencement of normal operation.

E.2.4 Design–build–operate contracts

E.2.4.1 Preliminary design

Addressing risk at this stage of the design process would be the same as for design–build systems. One advantage of this delivery form is mitigation of the concerns stated previously for the design–build contract. The designer–builder–operator will now be required to operate and comply with effluent criteria as well as pay for costs associated with operating the plant and maintaining the equipment. The benefit of this type of contract include transfer of most of the project risk to the contractor and having them responsible for the operations allows that bidder to develop what they feel is an optimum balance of risk and cost of the project.

E.2.4.2 Detailed design and construction

Most major DBO contractors have sophisticated risk analysis tools that can be used to balance the initial capital costs versus the predicted operational costs later. These tools range from as simple as applying contingencies and safety factors based on experience, to full Monte-Carlo risk analysis tools that can be used to estimate the cost impacts of various approaches. The tools consider process risks, design risks, construction risks, and operational risks versus the likelihood of their occurrence and the probable costs of the risks.

E.2.4.3 Operation

The design–build–operate (DBO) contract has significant risks related to long-term operational costs for the contractor, as compared to the owner doing operations. The DBO contract stipulates a cost for given loading conditions and what happens if conditions change, so the contractor is fully responsible for operation of the plant within the contracted loading and effluent conditions. Owners typically have more flexibility in their budgets for meeting changing conditions, within certain limits. The DBO does not eliminate cost risk to the owner should loading or effluent conditions vary from the contracted values.

E.3 CURRENT PRACTICE IN OTHER COUNTRIES

E.3.1 Questionnaire

The approaches engineers take during design vary depending on their geographic location. This is a result of the varying regional water situation and the legislative and contractual environment. In the following section, the approaches used by engineers in a selected number of countries across the world is presented. A questionnaire was sent to practicing engineers that included questions covering a variety of topics that capture the way designs are approached and the way risk is apportioned and handled. The questionnaire included the following questions related to design and risk:

(1) What is the prevalent type of contract delivery mechanism (Design (D), Design–Build (DB), Design–Build–Operate (DBO), Design–Build–Own–Operate (DBOO); if several, give percentages)?
(2) What are the most common types of design projects (green-field, replacement of an entire plant, upgrade of plant; give percentage range of capital cost of plant being replaced, give percentages for the three categories)?
(3) What information is included in the 'Requests for proposal' (RFP) prepared by the client (load projections, effluent requirements, configuration, industry standards to be used (e.g., ATV-131, 2000)? How much is predetermined, using which type of methods and which information (e.g., city master plans)?
(4) What is the typical design strategy of the engineering consultants (guidelines vs. mechanistic models, steady state vs. dynamic, calibration of models, performing of additional experiments on-site, safety factors used or parameter sets in models)?
(5) What is typically the design level at submission (level of completion of process design)?
(6) What are the typical bid selection criteria (including weighting) (e.g., cost, technical merit, ...)?
(7) Are post-audits performed (how is success/failure of a design defined)?
(8) What is the way that risk is typically apportioned (insurance – risk premiums)?

E.3.2 United Kingdom

Question	Response
Type of contract delivery mechanism	D DB
Type of design	*Anglian:* green field (<5%); new plant (<10%); upgrades (85%) *United Utilities:* green field (<2%); new plant (<5%); upgrades (>90%) *Severn Trent:* green field (<2%); new plant (<1%); upgrades (>95%)
Requests for proposal as prepared by the client	Internal design guidelines (alternative design guidelines implemented on occasion). Basis for design: flow and concentration Several of the water companies prepare designs internally.
Design strategy of engineering consultant	Team-based approach: engineering options listed and then eliminated. Remaining options investigated and proposals prepared. Presentation to programme board: questions asked, costs presented and process options discussed. Programme board approves proposal, or sends it back for more study. Water company and consultants work as a team start-to-finish on proposal. No bidding process: the consultants are part of a framework agreement, so RFPs as typically done in USA are not really applicable. The contractor agrees to provide an upgrade for a fixed amount of money. If the design runs over budget, then the water company will attempt to save the money on a different project. Costs debated internally. Hydraulics are always modelled (sometimes even with a physical model). Activated sludge plants might be modelled, but not always (steady-state typically, sometime dynamic). Safety factors: not specifically used, rather the design is sized to give an effluent concentration that is some percentage of the requirement (e.g., for a 10 mg/L effluent consent, the design will be sized to give an effluent of 3 mg/L). Conservative design parameters are used (e.g., conservative SVI, OTE, high max flow and loads). Pilot plants used for biological phosphorus removal trials and new innovative technologies.
Design level at submission	See above. Same team structure is used from project start to finish.
Selection criteria in bidding procedure	Capital cost Total life cost evaluated secondarily (Capital + Operational) For the Framework agreement: Contractor experience, financial stability, health and safety record and reputation.
Post-audit	Very little. Lessons learnt are incorporated into future projects. *Severn Trent* allocates budget to post project activities. Models used are not typically re-checked.
Risk spreading	Risk shared contractually with the contractors but in reality, the water companies assume the ultimate risk (contractor could be sued if negligent issue with delivery, design, etc.)

E.3.3 The Netherlands

Question	Response
Type of contract delivery mechanism	D (87%) New, more innovative contracts (risk-based approach, more risk in contracts, let market define solution). Recently some projects were done with 'innovative bidding', that was solution free (no design, only the problem was submitted). This appeared to put (too) much risk to the market and also led to non-optimal designs and a poor cooperation during construction (due to legal issues). Currently, there is a tendency towards working with framework contracts. In these contracts, a party or consortium is selected to cooperate and organise the different project phases, from performance/design specs to pre-design and construction.
Type of design	Plant upgrades (biology and secondary clarifiers).
Requests for proposal as prepared by the client	Design requested from 3 to 5 companies Design specifications: load projection and effluent requirements Standards: STOWA Guidelines on N-removal (HAS method), P-removal (Scheer method), bulking sludge guidelines, final clarifier design guidelines
Design strategy of engineering consultant	Activated sludge part design based on the ATV-131 (2000) guideline. The design approach includes: – Simple mechanistic model for nitrogen removal – Steady-state simulations – Influent fractionation (focus on determining biodegradable COD and volatile fatty acids, the latter for the Scheer Bio-P model) – Influent loading targets
Design level at submission	Preliminary design for the selection of the consultant and the design Detailed design for the final submission
Selection criteria in bidding procedure	Best elements from each of the four () preliminary designs. Detailed design. Tender for the construction. Quality and cost
Post-audit	Performance within the defined effluent requirements. Evaluation period. Technological evaluation.
Risk spreading	Typically, none, however, new design delivery methods are currently being tested.

E.3.4 Switzerland

Question	Response
Type of contract delivery mechanism	D: 90% (Assumption) DB: 10% (Assumption)
Type of design	Upgrades (biology, secondary clarifiers): 50% Re-dimensioning of primary clarifiers: 50%
Requests for proposal as prepared by the client	Effluent requirements, seldom load projections. No industry standards like ATV-131 (2000), but an orientation towards standards is desirable. State-of-the-art technology expected, but effluent requirements do not orient themselves on the best available technology.
Design strategy of engineering consultant	Mechanistic models: often combined with experience and guidelines. Steady state Safety factors
Design level at submission	Competition at the design stage very seldom (e.g., WWTP Bern or Zurich). Competition for building is price based. Winner usually proposes some modification to the original design during construction.
Selection criteria	Investment cost Technical merit Flexibility for further adaptations Yearly costs (includes personal, operation costs and value conservation) Acceptance by population (e.g., sludge treatment).
Post-audit	Compliance with effluent criteria after one year of operation
Risk spreading	None

E.3.5 Czech Republic

Question	Response
Type of contract delivery mechanism	D: 80%, DB:20% The design is almost always undertaken by a 'Project Company' which composes the design and project drawings. Some Project Companies may use an external expert/consultant for design. Together they compose the project management team. The Construction Company (usually the tender winner) must take on the project, check it and take responsibility for the delivery, including the guaranteed performance/effluent parameters. In rare occasions, the Project Company is the tender winner and will subcontract a Construction Company. In smaller projects where technology cost is high, the Technology Delivery Company may win the tender instead of a Construction Company.
Type of design	Small WWTP <2000 PE – 95% greenfield Medium WWTP 2000 – 100 000 PE – 99% upgrade Large WWTP >100 000 PE – 100% upgrade
Requests for proposal as prepared by the client	RFPs include load projections and effluent requirements. The design must comply with city master plans. No restriction is placed on technology selection however, the design must comply with best available technology given by legislation, the effluent standards and the Czech norms.
Design strategy of engineering consultant	ATV-131 (2000) guideline are often used. Compliance with Czech norms is required (norms include parameters such as load, SRT, HRT in reactors, etc.). Mechanistic models are used mostly in steady state. Dynamics are accounted for by applying 'irregularity coefficients' (e.g., for aeration system design). Irregularity coefficients reflect and correct for the real dynamic behaviour of the plant. Czech norms include several. For example: – hourly, daily, weekly, monthly irregularity coefficients – oxygenation capacity irregularity coefficient (different for small, medium and large WWTP can substitute a dynamic model with daily flow and load fluctuations). Standard modelling procedure: (1) Model calibration if data are available (if not, conservative values are used). On-site experiments are rarely required for model calibration. (2) Steady-state design (of several alternatives) leads to selection of one final alternatives. Steady-state design usually performed with more conservative parameters than what was used in the calibration model. (3) Dynamic design (parameters from calibration used). For industrial plants, lab or pilot-scale experiments are performed (in 10–20% of cases). Applying new technologies usually requires pilot testing.
Design level at submission	Tender documentation with basic plant design needed for cost calculation (technology information, basic process design, configuration, volume, air design, reactor depth, etc.)

Question	Response
Selection criteria	Merit: Compliance with effluent standards, city plan, ... Cost (90% weight) References, guaranties (5% weight) Technical equipment quality (5% weight) Environmental criteria have been newly (2021) added to the selection process however, there is no experience yet on how they will be implemented.
Post-audit	For projects which are partially funded from external sources, local, state or EU funds (grant-in-aid projects), there often (not always) exist additional criteria that must be met after the plant is finished. For example, the load capacity reached at a specific time. Grant-in-aid projects are audited after a trial period (usually 1 year). Plant capacity must be justified by proving that the plant is loaded to certain percentage of design capacity (usually 80%) and must comply with effluent standards.
Risk spreading	Client requires insurance of the contracted Co., together with other usual conditions for contract (financial, technical equipment, experiences, etc.).

E.3.6 South Korea

Question	Response
Type of contract	Turn-Key: 30–50% Separate contracts for DB, then operated by local governments: 50–70%
Type of design	Greenfield: about 30% Replacement of an entire plant: almost 0% Upgrades: 70%, as nutrients effluent quality is getting tougher
Requests for proposal as prepared by the client	Only load projections and effluent requirements.
Design strategy of engineering consultant	Design guidelines with safety factors.
Design level at submission	Complete process design
Selection criteria	Cost: 30% Technical merit: 60% Company's status (financial, previous records, stability...)
Post-audit	One- or two-years' successful operation (meeting effluent requirements).
Risk spreading	No insurance

E.3.7 South America

Question	Response
Type of contract	This varies enormously between countries. For example, municipal treatment plants in Brazil (run by government operators) are most likely D, whereas municipal treatment plants in Chile (being privatised) are mostly DBO. It also varies from municipal to industrial/mining/etc. treatment plants.
Type of design	Industrial treatment plant projects are evenly distributed between green-field and plant upgrades For municipal plants, there are quite a few greenfield sites in countries with low coverage (Peru, Ecuador). In countries with high coverage (Chile) plants are mostly upgrades.
Requests for proposal as prepared by the client	Varies by country. Projects funded by international financial institution, e.g., the World Bank, generally have load projections and effluent requirements (mostly concentrations). Local design standards are often cited in the request for proposals. Very few indicate design guidelines or standards to be used (e.g., ATV-131, 2000).
Design strategy of engineering consultant	Established consulting firms use dynamic modelling. However, there is a small number of these large, established firms. Most consultants use Excel spreadsheets with the Metcalf & Eddy steady-state equations. For larger plants (over 1 m^3/s capacity), established consulting firms tend to be hired. For smaller plants, Excel-based design prevails.
Design level at submission	Varies by country. Turn-key dominant in industrial applications, that is, the client 'buys' a plant from an equipment supplier that sells them the whole package (similar to a DBOT). Municipal plants are tendered with a preliminary process design. Bidders complete process design and execute detail design, construction and initial operation.
Selection criteria	Technical merit and cost. However, the primary selection criterion is cost. A few of the large utilities (two or three) are starting to give more weight to technical merit. Industrial clients: no defined standards, but mostly they pick companies they trust to work with and then look for lowest price. Little attention paid to life-cycle costs (present value including CapEx and OpEx). CapEx only dominates in both the municipal and industrial markets.
Post-audit	Rarely done. If design achieves quality criteria at start-up, then it is considered successful. Testing and auditing are not required. Some effort is put into evaluating the performance of the electrical and mechanical equipment.
Risk spreading	Through insurance.

REFERENCES

ASTM E2516-11. (2011). Standard Classification for Cost Estimate Classification System. ASTM International, West Conshohocken, PA, USA.

ATV A-131 ATV-DVWK. (2000). *A 131E Dimensioning of Single-Stage Activated Sludge Plants. German Association* for *Water, Wastewater* and *Waste*. GFA Publishing Company of ATV-DVWK, Hennef, Germany, ISBN: 3-935669-82-8.

Grady L. Jr, Daigger G. T., Love N. G. and Filipe C. D. M. (2011). Biological Wastewater Treatment, 3rd edn. CRC Press. ISBN 9780849396793.Great Lakes Upper Mississippi River Board (GLUMRB). (2004). Recommended Standards for Wastewater Treatment Facilities (Ten States Standards). Health Research, Inc., Albany, NY, USA.

Great Lakes Upper Mississippi River Board (GLUMRB) (2014). Recommended Standards for Wastewater Treatment Facilities (Ten States Standards). Health Research, Inc. Albany, NY, USA.

Metcalf & Eddy Inc., Tchobanoglous G., Burton F.L., Tsuchihashi R. and Stensel H.D. (2013). Wastewater Engineering: Treatment and Resource Recovery, 5th ed. New York, NY: McGraw-Hill Professional.

Melcer H., Dold P., Jones R., Bye C., Takacs I., Stensel D., Wilson W., Sun P. and Bury S. (2003). Methods for Wastewater Characterization in Activated Sludge Modeling, Water Environment Research Foundation, 99-WWF-3, IWA Publishing, London, UK, p. 596.

Tchobanoglous G., Burton F. L and Stensel H. D. (2003). Metcalf & Eddy Wastewater Engineering: Treatment and Reuse, International edn. McGraw-Hill, New York, NY, USA.

United States Environmental Protection Agency (USEPA). (1987). Phosphorus Removal Design Manual, EPA 625-1-87-001, Cincinnati, OH, USA.

United States Environmental Protection Agency (USEPA). (1993). Nitrogen Control Manual, EPA 625-R-93-010, EPA, Office of Water, Washington, DC, USA.

Virginia DEQ. (2008). Virginia's Sewage Collection and Treatment Regulations (form 9VAC25-790).

Wahlberg E. (2004). WERF/CRTC Protocols for Evaluating Secondary Clarifier Performance. Water Environment Research Foundation, Alexandria, VA, USA.

WEF MOP-29. (2005). Biological Nutrient Removal (BNR) Operation in Wastewater Treatment Plants, Manual of Practice 29. Water Environment Federation, Alexandria, VA, USA.

WEF MOP-8. (2009). Design of Municipal Wastewater Treatment Plants, Manual of Practice (MOP) No 8. Water Environment Federation, Alexandria, VA, USA.

Index

IWA Publishing's authorised EU representative for General Product Safety Regulations is Diane D'Arras, 15 rue Duret, 75116 Paris, France, e-mail: safety@iwap.co.uk.

Printed and bound by CPI Group (UK) Ltd, Croydon, CR0 4YY
16/06/2026
02140399-0001